Praise for *Client Earth*

"When compassion for life on Earth, or call it fury at the everyday and casual destruction of our stunning biosphere – when these two powers are honed by these exceptional environmental lawyers to a fierce and fine point, change happens, and the world is a better place for it. Humanity's grace and dignity are restored each time a case is successfully brought and won – it is truly a global battle fought between the mindless proponents of tragedy and the (sometimes extremely) courageous proponents of hope."

—Emily Young

"ClientEarth have been pioneers in using the tool of environmental legislation to tackle the modern scourge of air pollution. This is the story of how they're doing it."

—Ed Miliband

"The story of ClientEarth – and of its charismatic Founder, James Thornton – is truly inspirational. His only client is our battered, abused planet, and his favoured arsenal is the rule of law in defence of public interest. The hard-fought victories that you'll hear about are all important, but more important still are the vision, values, and gritty dedication of an amazing group of lawyers and campaigners to whom we owe a very great deal."

—Jonathon Porritt

"This book is an inspiration for those of us trying to build a sustainable future – and I'd recommend it to anyone who wants to know how and why we must deploy and enforce the law in the fight against ecological destruction."

—Caroline Lucas

"The foreword by Brian Eno is an eye-opener in itself: well-written, well-argued and inspirational. This book is different from any I've read and was thus very stimulating. It is a good read and the messages are put across with true stories that are well told. I recommend it as a thoughtful read. And it may be that it carries messages that are particularly pertinent to our times and our predicament."

—Mark Avery

Client Earth

Martin Goodman is the author of nine books of fiction and nonfiction. His most recent book of nonfiction, *Suffer & Survive*, won First Prize, Basis of Medicine in the BMA Book Awards. He holds the chair of Creative Writing at the University of Hull, where he is director of the Philip Larkin Centre for Poetry and Creative Writing.

www.martingoodman.com @MartinGoodman2

James Thornton is founder and CEO of ClientEarth, the only pan-European group of environmental lawyers working in the public interest, throughout the EU and in Africa, China, and the US. He is a solicitor of England and Wales, and a member of the bar in New York, California, and the Supreme Court of the United States. The *New Statesman* named him as one of ten people who could change the world. He is also a Zen Buddhist priest, a naturalist, an author, a violinist, and a birdwatcher. He lives in London and the Languedoc, with Martin Goodman.

www.jamesthornton.co.uk @JamesThorntonCE

Client Earth

Martin Goodman
and
James Thornton

with a foreword by Brian Eno

SCRIBE
Melbourne • London

Scribe Publications
18–20 Edward St, Brunswick, Victoria 3056, Australia
2 John Street, Clerkenwell, London, WC1N 2ES, United Kingdom

First published by Scribe 2017
This edition published 2018

Typeset in 11.5/16.75 pt Adobe Garamond by the publishers

Printed and bound in the UK by CPI Group (UK) Ltd, Croydon CR0 4YY

Scribe Publications is committed to the sustainable use of natural resources and the use of
paper products made responsibly from those resources.

9781947534032 (US edition)
9781925322040 (ANZ edition)
9781911344087 (UK edition)
9781925307993 (e-book)

A CiP record for this title is available from the National Library of Australia and the British
Library

This book was written with the support of the McIntosh Foundation,
Washington, DC

Some of the proceeds of this book will benefit ClientEarth
and will help fund its international legal work.

scribepublications.com
scribepublications.co.uk
scribepublications.com.au

For Michael McIntosh
1933–2015

with thanks for his generosity,
love of the planet, and spirit of the fight.

Contents

'Men argue, nature acts.'
Voltaire

*'A man cannot despair if he can imagine a better life,
and if he can enact something of its possibility.'*
Wendell Berry

Foreword

Brian Eno

At a certain point in life, if you're a certain kind of person, you find yourself thinking that you could use some of your capabilities to make the world a better place. And if you spend any time thinking about it, you'll probably want your contribution to have as big an effect as possible. You want to apply maximum leverage, so that your efforts can in the end translate into a meaningful result.

Like everybody of my generation, I grew up with the buzz of environmental catastrophe ever louder in my ears. I know — we all know — that things are looking bad, and that whatever was going wrong before is generally going wrong faster now. It feels like we're on an ever steeper slope, and about to reach the point where we can't hold on anymore — free fall.

So what to do? Well, there's always *do nothing* — you can just deny that it's happening, in the face of all the evidence. We humans have an astonishing capacity to ignore evidence that doesn't suit our interests and to amplify any scraps that do. The attraction of that approach is that everybody can continue with business as usual and make lots of money. But I'm reminded of that Tom Toro cartoon in *The New Yorker* where a dad is sitting with his three ragged children in a cave, a ruined city smoking in the distance. 'Yes,' he says, 'the planet got destroyed. But for a beautiful moment in time we created a lot of value for shareholders.'

Another form of doing nothing is to play the techno utopian card and insist that something will turn up to save us — for example, some amazing new way of generating energy so clean and efficient that we could use it not only to power all our industries but also to suck excess CO_2 out of the atmosphere. It's a great idea — but what if it doesn't turn up? Of course, we should vigorously pursue every technological fix possible, but to put all our eggs into that optimistic basket requires a leap of faith that I suspect few of us would be willing to make. And anyway, it's not a solution but a postponement. Something more fundamental is required.

So if you decide that doing nothing isn't for you, what do you actually do? Simply mopping up the results of environmental crises, though essential, doesn't seem enough — it would be better if the oil never got spilled in the first place than that we end up scraping it off the wings of gulls, or better that sea levels stop rising than that we deal with the vast waves of displaced migrants when they do. By the time things have reached crisis point — which is approximately the moment that our governments and media will notice them — it's too late. The worst has happened. We have to get in before that, to penetrate these processes much earlier in their development.

To do that effectively means playing with a deeper set of levers, changing the ground rules to the extent that certain behaviours become socially impossible. As Paul Steinberg says: 'If you want to change the world, change the rules. A rule is just an idea with an anchor attached to it.'

That's really what ClientEarth is about — to link the idea ('We need to change our behaviour!') to the anchor of law ('And now we have to!'). Law is the most definite type of rule that we make: the expression of democratic consensus backed by the power of the state.

Law works best when it makes sense to people, when it is seen to support and defend a moral consensus. Nobody has to defend theft laws, for example: we intuitively grasp the logic of private property. We need to cultivate the same sensitivity to public, common property. We need to make it clear that care for the environment is a central *moral* issue. And I think we can.

Humans are malleable, and born with strikingly few intrinsic moral constraints. I'm sure if you gave a child licence to eat human flesh from an early enough age, it would seem as natural to her as eating carrots. If you encouraged stealing or bullying or promiscuity or cheating, I imagine you'd have a good chance of producing a society of thieves, bullies, and cheats. It's our cultures that impress upon us that some forms of behaviour aren't acceptable, that make us, by and large, non-thieves, non-bullies, and non-cheats. Standards of appropriate behaviour can be passed on through customs and taboos — but the strongest and most enforceable form in which they are communicated is through laws.

By insisting that certain behaviours are illegal — and punishable — we make a firm and clear statement about how society regards them. Such statements are usually resisted at first, but become 'common sense' in time. Think of slavery: it seemed natural to 18th century Europeans, but very shortly after it became illegal it also became unthinkable. Think of universal suffrage: a totally ludicrous notion to most people until it happened. Think of child labour … and even of smoking bans in pubs! Who even argues about those things now? All those deeply felt opinions changed, and the changes were often surprisingly swift. As Alexei Yurchak said about the disappearance of the Soviet Union, 'Everything was forever, until it was no more.'

We are now at a point where we need to make a similar leap, to really grasp that our planet is finite, exhaustible, and vulnerable. Ninety nine per cent of our historical human experience is of an inexhaustible planet with abundant resources. We haven't got used to the idea that we could use it all up. We need urgently to make a transformation in our attitude — at the personal level, at the commercial level, and at the level of government.

Can we engender a situation where environmental irresponsibility is as unacceptable as, say, child labour? The problem is that we are still in the grip of a politics which, even if it admits the possibility of environmental degradation, assumes that's an acceptable cost for our wellbeing now.

But who is 'us', and when is 'now'? If 'us' is to include all the inhabitants of the planet, then we have to start rethinking our attitude to, for example, dumping the waste we don't like onto developing countries. And as for 'now' — if our calculations of the costs and returns of a behaviour extend only into the next year, they will likely be different than if they extend into the next century. At this point, we're trapped in very small time frames — quarterly earnings reports, GDP, high velocity trading — which defer the actual costs of our behaviours into a future we aren't thinking about. We are in the lazy habit of externalising the costs of our behaviour to some other place, some other time. It may have worked once, but it isn't working now.

We should be thinking of ourselves as a global community with attendant responsibilities, not as a collection of individuals engaged in a Darwinian struggle at any cost to the planet. What does 'individual' mean anyway now, when almost everything we do — like taking a train or buying the groceries — involves us in the accumulated brainpower of thousands of other people who've set these systems up and maintain them? We survive as individuals only because we know how to access the brainpower of millions of others. We aren't any longer lonely nomadic heroes roaming the landscape: we're an intensely interdependent global community of highly specialised brains. I'm a musician. James is a lawyer. My daughter's a doctor. We're all specialists who command the power of thousands of other brains — and who probably wouldn't survive the week without them. This is a human transition we could celebrate instead of denying.

But we're stuck in the atomised, individualistic picture we inherited from the Age of Reason. Nearly half a century down the line from those hyper individualists Thatcher and Reagan, we see a world in disarray — ever increasing inequalities of wealth and opportunity; environmental chaos looming; governments losing the consent of their discontented populations and clinging to power by increased surveillance, propaganda, fearmongering, and coercion. The world of unrestrained individualism turns out to be one that most of us don't want to live in.

We have to discredit one set of attitudes and dignify a new set. We have to be able to express our sense of community over our sense of difference. Isn't this what democracy is supposed to be about — the attempt to benefit from the wealth of different intelligences? But how can this happen when it seems the whole democratic process is broken, when the wealthy — whether in human or corporate form — can buy the politics they like and skip off to another jurisdiction when it suits them? How do we persuade the winners in the system that they can't continue privatising their profits and socialising their costs? How do we reinstall a sense of responsibility? We can do it with campaigns, drawing attention to the worst players. We can attack them in the press. We have to do that, but, as George Monbiot remarked, 'The pen might be mightier than the sword, but the purse is mightier than the pen.'

But that's exactly the clue. As James Thornton says, 'Money is the grammar of commerce.' Law can *trump the purse* by levying fines and placing limitations on the freedom of states and businesses to behave irresponsibly. In the business world, that's the strongest sanction possible, the one that really counts. In the absence of strong legislation backed by serious penalties, the current system favours the cheats. If it costs money to clean up your act, and there's no real penalty for not doing so, then the playing field is tilted towards those who cheat the system. There are many good and conscientious people in business and politics, but currently the system is stacked against them — effectively, they are penalised for acting responsibly as long as their badly behaving competitors are getting away with it. Irresponsible businesses and governments need to be faced by two realities:

- That people object to them so strongly that they stop buying their products or voting for them.
- That the law makes life so difficult for them that they have to change their behaviour.

To the first: We need a social revolution — a revolution in public opinion — that makes bad behaviour too disgraceful, too shameful, for a company or a state to countenance it. We need to name and shame. This is beginning to happen now in the wake of the 2008 banking crisis and the tax evasions routinely practised by many big multinational companies. People are alert to those practices and find them offensive — cheap and tawdry. This public disapproval will translate into something — though it could just be greater subterfuge unless it's accompanied by the force of law … which is where ClientEarth comes in.

I remember James telling me some years ago that there were 18,000 industry lobbyists at work in Brussels, pressuring the European Union to draft its laws in their favour. Speaking up for the environment was a tiny handful of people — ClientEarth among them. On the face of it, those odds aren't good. All that we have on our side of the argument is the brainpower and commitment of people like James to tilt the rudder in a new direction, and the power of public opinion to support it.

It can work. It is already working, as you will read in this book. Once something becomes law, it becomes actionable and enforceable. Shortly afterwards, it becomes 'common sense'. It becomes possible for a small group of people like ClientEarth to use existing legal structures to trigger big changes. The task is pretty daunting, but the rewards are huge, for if we achieve the escape velocity necessary to catapult ourselves out of the mess we're in now, who knows how far we'll go. Dealing with these thorny environmental issues will require global human cooperation on a scale that has never been seen before. If we can pull it off, we will be on the way to saving not only the planet but global civilisation too.

Disclosure: I've been a trustee of ClientEarth for eight years. This book has delighted me with its wonderful story of how ClientEarth came into being. I feel proud to be a very small part of this organisation; proud of Winsome and Michael McIntosh, who planted and tended the seed; proud of James and the staff for keeping it growing and making it so effective; and grateful for the feeling that it has given me the chance to do something a little useful in my life.

Introduction

Martin Goodman

A new breed of lawyers act like sheriffs in a previously lawless land. They have the badges while marauders with six-shooters roam the plains. Instead of weapons, these new sheriffs wield sub-clauses in statutes that no one much cares for. Theirs is an unappreciated and hazardous way of life.

Law offers remedies that can stave off ecological collapse. However, laws are meaningless when not enforced, and a main thrust of this book is the story of enforcement, including the writing of laws and regulations that are enforceable. You cannot enforce invisible laws, and so it is important that readers come to know the laws that protect them. Most of the environmental laws that are effective today are new. Laws are as vulnerable as species, and need their own safeguarding, so it is vital that all of us both know about them and value them.

I learned to value the work of these lawyers, and the fabric of laws that support life on Earth, by wandering in as an outsider. James Thornton and I have shared a life for 25 years, so I have had a clear view of his environmental work from the domestic front. To tell the whole story, I had to muster critical detachment and enter the field. Chapters are in my voice as I lead you on that journey. James's essays draw from his experience as a frontline public interest environmental lawyer and interleave the chapters. They broaden into a moral dimension. I choose to say moral rather than philosophical, because moral combines

1

deep reflection with action. The chief effect of this book is that it shows how rules govern behaviour, and provides evidence based solutions to ecological crises.

It was early morning on a stretch of Southern California coastline.

Bright with sunshine and studded by palm trees, Santa Barbara is as tidy and clean as a Disney vision. Dowitchers and redshanks patter after the retreating waves, their beaks stabbing the sand for food. Brown pelicans fly low lines above the sea, and stand beside the anglers on the pier. Oil platforms are visible from the shore, and beyond those lie the humps of the Channel Islands. Further still, grey whales ply the route between the feeding grounds off Alaska and their breeding sites in Mexico's Gulf of California.

James is alert to every nuance of nature, unless he's thinking. Today, he was thinking hard. We had walked the beach down as far as the cliffs at one end, and now we were almost back to the pier. Public interest environmental law has some good recent history in America, and James was at the forefront of that. Now, we had relocated to London, where he received a challenge. Bring that legal aspect of the environmental movement to Europe. What should he call such a group?

Names had come and gone. They all seemed lame, or ponderous, or already taken. 'You've often told me who your client is,' I reminded him. 'Your client is the Earth. That's what you say.'

He looked at me. What was I getting at?

'There's your name,' I said. 'ClientEarth.'

He gave me a maybe kind of smile. The name was interesting, but maybe too quirky for a serious law group?

He looked out to sea. A small commotion was happening in the waves that lapped the beach beneath the pier. James stepped closer. A pigeon sat there. A white neck and head reached out of its grey body towards its grey crown. It blinked its eyes and shivered its wings. It was dying yet painfully alive. A herring gull was stabbing its beak into the pigeon's head and body, snatching mouthfuls of living flesh. These gulls nest out on the Channel Islands, and fly across the ocean to scavenge the land. James clapped his hands and shouted. The gull backed off a few yards, but waited.

James picked the pigeon out of the water and wrapped it in his hands. The bird's heart beat fast and then calmed. The gull waited. James's eyes moistened. He breathed in, took the pigeon's neck in his right hand, and twisted. It was death — tough on James but kinder for the pigeon than being eaten alive. James set its body down. Gulls scuttled in to feed on the corpse.

We're on a journey into nature, but it's not a sentimental one.

We headed for breakfast of our own, tucking into eggs at a diner and flipping through possible names. Back in the motel, we ran those names through Google. All were taken. ClientEarth? It brought zero hits.

The name stuck. In years to come, James would look down from his ClientEarth office. Plate glass windows faced straight onto the playground of London Fields. Squirrels climbed the trees, and so did children. More children slipped down the slide, swung on the swings, screaming and laughing. 'Those are my clients,' he explained to visitors.

James is one of a lawyer's four sons who all became lawyers. That has a ring of inevitability about it, but for James the choice was simply practical. As a youngster, he saw that the planet was in ecological crisis.

James went on to win top honours in Philosophy at Yale. He learned why humans might treat the planet in the way that they did, and to see how they might build an egocentric worldview. Philosophy offered no obvious remedies to ecological devastation. To bring about meaningful change, he needed a set of tools. He judged that study of the law would provide him with the best set of such tools.

What is the basic make-up for a crusading lawyer? Imagine them at the edge of the school playground, blinking with wonder at the human predicament. Theirs is not the flash response. Instead, they observe and they analyse. They stare out at the world, and question how inequities can exist. And then they seek a remedy. This book charts something of the phenomenon of the environmental law group, from its conception in the USA to its arrival into Europe and global spread. It is in such a home that crusading lawyers make their base.

These lawyers have the same emotional response to environmental horrors as those brave souls who chug out in inflatable dinghies to counter whale fishing fleets, or those activist campaigners who climb smokestacks or invade oil rigs. It's just that they redirect the thrust of their emotional response. Anger turns them quieter. It sets their brains working. Some environmental activists claim that such lawyers are not activists at all, and indeed the lawyers often reject the terms environmentalist and activist for themselves. Theirs is evidence based activism. It's slow burning. It can mean self-sacrifice and sometimes danger. It involves threat without violence, and aims at resolution.

Humans have always used the Earth as a resource. Think of the Earth as a pie. Traditionally, lawyers join the well paid squabble over how the pie is divided. These crusading lawyers accept much lower wages in order to work out what is best for the pie. The Earth is their client. They tune in to what the Earth needs in order to remain intact, and then they work through the agency of law to ensure that happens. When they get it right, gains for the Earth are their end of year bonuses.

Our future on the planet depends on their success.

This book suggests a model of public interest environmental law that began in America and then was brought to Europe by Americans. That's a significant postcolonial act — the 'Old World' is being taught aspects of environmental regard by the 'New World'. What might be so particular about American legal environmentalism?

From among those who settled the East Coast in colonial times, some pioneers moved west. They found a land of Native Americans, who lived close to nature; grizzlies pawed salmon from rivers, bison thundered across the prairies, bald eagles filled the branches of tall trees. These pioneers wrote in wonder of the natural beauties that they found, and recognised the environmental impact of their incursions. The world's first National Park was created at Yellowstone in 1872. It was a great act of conservation, but it had a flipside. You conserve something because it is threatened. Yellowstone was protected, which meant the rest of natural America was fair game. At the beginning of the 20th century, children could watch passenger pigeons blacken the skies in their millions. Mid 20th century children never glimpsed the species. It was extinct.

When your direct ancestors walked your country and it was like an Eden, it lodges as a fresh kind of pain.

In October 2016, I ditched my charity store dinner jacket and invested in a new tuxedo. James was up for two awards in the space of a week and it was time to look the part.

Europe's top lawyers pay £600 a ticket to attend the annual *Financial Times* European Innovative Lawyer Awards. Dining tables spread across the vast entrance hall of London's Natural History Museum, beneath the dinosaur skeleton of the diplodocus. Our table was at the front. James was set to give the keynote speech, and receive the Special Achievement Award.

I scanned the programme. 'Did you know ClientEarth is up for its own award?' I asked James. 'FT 50: Most Innovative Law Firms 2016.' This was news. Before long, James and two of his team were on stage and clutching the award, with smiles for the photographers. For his words of acceptance, James thanked the philanthropists who made his firm possible. They are big players in the story you will read, canny folk who use their wealth to save the planet.

I had smiled when I read James's keynote speech for the night. It compacted the nine years of ClientEarth's existence into eight minutes of text composed for legal minds, a dense yet lucid unfolding of intelligence. It would lose an average audience. How would top lawyers cope on their big night out?

James began. Party chatter still rose from distant tables, and then a wave of silent attention reached back to enclose everyone. Once he'd finished, to a rush of applause, James took his award and sat down, glad for a drink at last.[1]

It took an email some days later to bring home the true significance of the night. The message came from Pierre Kirch, competition lawyer and litigating partner in the Paris and Brussels offices of the top law firm Paul Hastings, and an adviser to ClientEarth. 'Special Achievement is amongst the ten award categories, perhaps the most prestigious,' he wrote. 'Of the other nine categories, ClientEarth was the hands-on leader of the five law firms cited as the "Standout" firms in "New Models". The citation merits mention, as it states in concise terms who we are: "ClientEarth is Europe's first law firm set up to defend the public interest in the environment. Formed as a charity and independently funded, it now has 100 staff, who use the law as a strategic tool to protect the environment and human health … The firm aims to change legislation and policies and influence how businesses report their impact on climate. An important part of the firm's work includes making sure that environmental laws are implemented in the spirit of how they were written."'[2]

Pierre went on to read down the list of FT 50: Most Innovative Law Firms 2016. 'Worldwide players that have been there forever, such as Linklaters, Allen & Overy, Hogan Lovells, Baker & McKenzie, Freshfields, Shearman & Sterling, Paul Hastings, Skadden Arps: these firms are amongst the top 25 on the list,' he noted, 'but who comes in at number 46? ClientEarth. This is based on a global score across all categories, most of which ClientEarth cannot compete in. It goes without saying that it is alone in the list of the top 50 as a "charity" law firm, and this, moreover, as a firm that is not specifically specialized in the business of innovation in the legal sphere but specialized rather in the use of the legal tool to protect the Earth.'

I read Pierre's comments and realised that this book is an ugly duckling story: the poor relation charity environmental law group that suddenly found itself among the swans of top global law firms. It is a tale of a quiet revolution within the environmental movement. We know that the side with the best lawyers tends to win. Now, as the 21st century sees us in a period of intense ecological crisis, the Earth has the best lawyers.

America is the land of Marvel superheroes. We are introducing a different style of hero in this book, so we'll reverse and subvert the comic book story a little:

Superwoman beams in from another planet. That gives her planetary vision. She sees the Earth will be devastated if it carries on along its present track. As a superhero, she has to save it from disaster.

It takes a while, but she finds an old phone box with a door that closes. She strips out of her tights, tears the self-aggrandising logo from her chest, and folds up her cloak. They're no use for a while. She could clench her hands into a fist and rocket through the atmosphere and tackle crisis after crisis, but the world doesn't need a Saviour. It needs to take responsibility for its own actions.

She puts on an open neck shirt, a dark suit, a pair of glasses, and emerges from the phone box as Lois Lane. Journalists are important, she'll need those to communicate her message, but she's just become a lawyer. She can use those glasses to scrutinise the fine print on legal documents. That's where humans have codified their sense of the right and proper order of life on Earth.

She marches the urban streets and enters the offices of a public interest environmental law group. Clark Kent's head faces a computer screen, he is talking into a miniature microphone, but even so he recognises her as she enters. Clark turns to Lois Lane and smiles. He's been at this stuff a while. As Lois settles down beside him, he works to bring her up to speed. They start local, looking at current environmental legislation at regional and national level, and then go global.

'Is that it?' Lois asks.

Clark nods. He looks rueful. 'Environmentally, we live in a lawless world,' he says.

Lois takes off her jacket and rolls up her sleeves. She's an enforcer. What laws there are, she's set to enforce them. Maybe she can clear away those obstacles that block entry to the courts, so that citizens can take action for themselves. Where laws don't exist, write new ones.

That's all clear, but she's a superhero in Lois Lane disguise. She needs a clear, mean, powerful enemy. She walks to the triple glazed window and looks out at a baking world. Air conditioning chills her back.

'Who's responsible for all this?' she asks.

A young legal intern sits beside the desk. 'Me,' she says. 'I am responsible. My suit was made by wage slaves in Bangladesh, and shipped here in containers. I received a scholarship to attend a university which was funded with the fruits of empire. Men died to mine the components in my mobile phone. I once flew to Thailand to go snorkelling. My hand cream contains polyethylene which sends billions of nanoparticles into the oceans whenever I wash. As a citizen of the developed world ...'

Lois puts up a hand to stop her. She does not need this. She needs public enemy number one.

'Here,' Clark says.

Photos of men in suits populate his screen.

'They support corporate actions that increase the dump of carbon dioxide into the planetary sinks. Their profits buy direct access to Heads of State.'

'Right!' Lois Lane pumps a clenched fist into the air. 'Let's sue the bastards!'

'Sue the bastards!' was an early slogan of crusading environmental lawyers in the USA back in the late 1960s. The movement soon became more professionalised and nuanced.[3] However, I offer the Superwoman tale as a trailer, because this book really does tell the story of people fighting an almighty battle to save the planet.

They can't do it alone. By reading this far, you are already joining in.

1

The Voice of Many Waters

A curve of land protects a stretch of coast below Washington, DC, from Atlantic storms. Chesapeake Bay is the largest estuary in the United States, with a surface of 9,920 square kilometres, which before the 19th century held 500 square kilometres of oyster reefs. Those oysters were able to filter the entire waters of the bay inside three and a half days.

Early sailors looked down through crystalline waters to watch the blue crabs some six metres below. Captain John Smith sailed into the bay in 1607, the first Englishman to see and map it, and wrote of the shores paved with oyster shells. It was clear that the 'Savages' fed off these plate sized creatures. Captain Smith and his crew planted a cross where a great river fed into the bay, and named the river after their English king, James. By language and symbolic action, they were staking the river as their own. A young English nobleman with the party studied the James River and marvelled at the 'great plenty of fish of all kinds. As for Sturgeon, all the World cannot be compared to it.'[1]

By the mid 19th century, an oyster industry had been established in Baltimore. The bay's oyster reefs were dredged to near obliteration during the late 19th century oyster wars.[2] As a child in the 1880s, the writer H.L. Mencken recalled that Chesapeake Bay was still 'an immense protein factory' from which Baltimore 'ate divinely'. This was despite 'the polluted waters of the Patapsco River, which stretched up fourteen miles from the bay to engulf the slops of the Baltimore canneries and fertilizer factories.'[3]

The bay holds roughly 75 trillion litres of water. By the 1980s, about a fifth of that was wastewater discharged from the industries and sewage plants of Maryland and Virginia, and 4.5 million tonnes of that was composed of four common pollutants. Because of the way the bay is formed, only 1 per cent of those pollutants were flushed out into the Atlantic. For a decade from the mid 1970s, the already depleted annual oyster catch was reduced by a further two thirds. Ninety per cent of the nation's rockfish came from the Chesapeake. The 2.7 million kilograms of rockfish caught in 1970 was down to 270,000 kilograms in 1983 when a ban and restrictions on commercial fishing were introduced.[4]

Such degradations of planetary ecosystems are commonplace, and already centuries old. Humans raided the Earth, and formed legal systems to help them retain the spoils. Huge scientific and philosophical endeavour recently recovered the truth that humans are not independent of the biosphere. That realisation is so radical that people absorb themselves in commentary about what it means. Meanwhile, new environmental laws came into effect that were not simply about people's property rights, and a small professional group formed to focus on deploying them.

Action is an antidote to despair, so it is good to learn of the actions that public interest environmental lawyers take to mitigate ecological damage. The work of these lawyers is evidence based. Their numbers are few, and the global imperative of such actions as stopping carbon emissions is so vast and immediate, they must continually escalate their effect. Their work has an organic nature about it; they investigate how systems work, and then root down to intervene at those points where health and harm remain possible outcomes.

The first test was this: could a lawyer reverse the industrial scale pollution of Chesapeake Bay?

There are tides in politics as well as in bays. One rise in political tide brought in a new law which could have helped save the bay, but when the tide ebbed the new law was rendered useless by neglect. (We'll see that there seems to be roughly a decade between these high and low tides, when the environment washes high on the political agenda and then away again.) Public interest environmental lawyers are needed at both points: when laws are getting written, they can help them be more robust, and when the laws are neglected they can enforce them.

The environment barely figured in the US presidential campaign that saw Richard Nixon elected to office in 1968. Public concern for the environment then saw a sharp and steep rise. In January 1969, pressure systems failed on an oil platform in the sea ten kilometres out from Santa Barbara. Within ten days, roughly 100,000 barrels of crude oil had washed towards the beaches of southern California and devastated a rich marine environment. News screens were swamped with the images. A visit to the scene inspired Senator Gaylord Nelson to set up Earth Day, with a staff of 85, cohering disparate citizen groups, all with separate interests. A proposed jetport in Miami threatened the Florida Everglades. A pipeline in Alaska was set to run across pristine landscape. Lake Erie was said to be 'dying'. Smog in the cities, phosphates in detergents, lead in petrol, fish contaminated by mercury, bald eagles threatened by DDT — these were some of the environmental problems that excited public concern.[5]

Earth Day, April 22, 1970, was set to be a 'teach in' on these issues, and Congress was forced to close for the day as so many politicians chose to take to the streets of their constituencies to show their support.

A poster on a wall of a high school classroom in South Bend, Indiana, showed a riverbed, cracked and dried, with a dead bird laid upon it. It called on young students to march with those Earth Day millions. James Thornton had just turned 16. He considered Earth Day, and it didn't appeal. It was a people thing, and he and popular didn't really get on. Whenever Notre Dame, the local football team, played, he would close himself in his room and play his violin for hours, so as to drown out their noise and live inside his own world of sound. Earth Day was too feel-good for him, too emotional a response to an urgent need. The Earth was in desperate need of practical action, not a group hug. He kept to his desk. There was plenty he needed to learn before he could be any use at all.

His studies were solid for ten years. A Philosophy degree from Yale included graduate level courses in the natural sciences, but still James did not feel equipped to face ecological crisis. He moved on to New York University, edited their Law Review, and emerged with what he had decided was the best skillset for mending the planet: a Juris Doctorate (JD) in Law from New York University.

Meanwhile, the United States under President Nixon had brought in a body of environmental laws to stock a lawyer's toolkit, 'an extraordinary environmental record in almost every respect and one that is certainly without parallel in any administration that has followed' according to experts.[6] Starting with the National Environmental Policy Act and the Clean Air Act of 1970, 'from 1970 through 1976, the environmental legal system as we know it today was largely established. This mosaic of environmental laws and implementing regulations provides the foundation for the environmental legal profession.'[7]

In 1970, Congress set up the Environmental Protection Agency (EPA) to help implement and enforce its new laws. The law that the Chesapeake needed came with the Clean Water Act of 1972, that set controls on industrial emissions into the nation's waters. Congress recognised the potential of political drift to destabilise the EPA in their enforcement role, and so the Clean Water Act allowed for citizens' suits whenever the EPA was seen as not performing its duty. 'By empowering citizen enforcers, Congress established a mechanism for "policing" the enforcement relationship between regulators and regulates.' Matthew Zinn notes the danger of 'regulatory capture', whereby the regulator is seen to be controlled by the industries it is meant to be regulating. 'The statutes provide the means and incentives for interested pro-regulatory groups to participate where capture is most dangerous: enforcement.'[8]

That environmental political tide had turned by 1981. The Administrator of the EPA is a presidential appointment. Anne Gorsuch recalled her interview for the post with the new President, Ronald

Reagan. 'Kind of quiet like, but dead serious, [he] asked, "Would you be prepared to bring EPA to its knees?"' Gorsuch was so startled she laughed. Reagan had found his woman.[9] Two years later, *The New York Times* noted that:

> Mrs. Gorsuch has undermined the E.P.A. by halving its budget when its responsibilities are doubling. She has induced many of its best professional staff to quit, and has sabotaged the agency's enforcement effort by continual reorganizations and cutbacks. She has scrimped on the science and monitoring that must underlie effective regulation.[10]

Gorsuch abolished separate enforcement offices within the EPA and reassigned their staff. EPA officials were told that recommendations to prosecute violators would be noted as 'black marks' against them. 'Agency morale, especially among personnel engaged in enforcement, reportedly declined as rapidly as the case referral statistics.'[11]

Public interest environmental lawyers in the USA grouped into their own organisations as the new environmental laws came into being, so that regulation was not simply in the hands of the government. One such group was the Natural Resources Defense Council (NRDC). Jeffrey G. Miller, then head of national enforcement for the EPA, was one of those forced out from that regulator by the breakdown in enforcement. He has written how national environmental organisations sought to reverse the trend. 'With a seed money grant to fund a few initial citizen suit enforcement cases, NRDC hoped to produce a self-sustaining effort by recovering attorneys' fees and using them to fund future cases.'[12]

That self-sustaining effort was named the Citizens' Enforcement Project, and it started off with a loan by the philanthropist Michael McIntosh. James had interned for NRDC, and was appointed to run the project. He was the sole attorney, supported by the chemist Patrick O'Malley in the role of project scientist, and with the engineer Bruce Bell as project consultant.

Its purpose?

James was to take over the government's role, defeat and punish the big industrial polluters, and so shame the government back into active law enforcement.

This at last was a call for practical action for which James felt equipped. He had his vocation, and he was unleashed. Working out of New York, his initial focus was on the Northeast with some action in the South and Midwest: suits filed in the states of Maine, New Hampshire, Massachusetts, Connecticut, Rhode Island, New York, Maryland, Virginia, Louisiana, Ohio, and Michigan. He often partnered with national and local environmental organisations, linking them into a national legal strategy. For the Chesapeake, he approached the Chesapeake Bay Foundation. Together, they targetted two principle polluters: Bethlehem Steel in Maryland and the meat packing company Gwaltney of Smithfield Ltd in Virginia.

Travel up the James River from where it feeds into the Chesapeake and you come to a 20 kilometre long tributary, the Pagan River. On its banks is the historic seaport of Smithfield. Heading towards the Smithfield Station marina, a recent pleasure sailor found the river 'hauntingly gorgeous'. He also noted 'a cloying odor of — how to put it delicately — rendered animal flesh'. He 'wondered, too, about the health of the Pagan River, which has been closed to shellfish harvesting for 30 years due to high levels of fecal bacteria'.[13]

In the early 1980s, 1,500,000 hogs a year made the one way journey to the Smithfield processing plant on the outskirts of the town, which markets itself as the Ham Capital of the World. Holding ponds, named 'lagoons' by the company, surround the plants and contain their flow of industrial pig waste. There can be hundreds of these lagoons around a slaughterhouse, some of them nine metres deep. Hogs produce three times more excrement than humans, so you would expect these lagoons to be brown. In fact, the excrement contains high doses of antibiotics, vaccines, and insecticides, which are used to keep the pigs alive in packed quarters. The waste contains methane of course, ammonia too, and cyanide, carbon monoxide, phosphorus, nitrates, hydrogen sulphide, and heavy metals. Throw in blood, urine, stillborn piglets that slip through the waste grilles, and then bacteria, which all join to turn the lagoons pink. That faecal bacteria that contaminated the shellfish? Well, just one gram of pig shit can contain 100 million faecal coliform bacteria.[14]

Slurry from the lagoons was used as fertiliser to spray surrounding fields. Meanwhile, for several decades, 'the Pagan had no living marsh grass, a tiny and toxic population of fish and shellfish and a half foot of noxious black mud coating its bed. The hulls of boats winched up out of the river bore inch-thick coats of greasy muck.'[15]

The Clean Water Act required industries to obtain permits to discharge pollutants into national waters. The permits set limits for effluents in different categories, and those with the permits were obliged to monitor their own discharges and log them. Comparing those monitoring reports with the limits reveals the violations. Filing a complaint about these violations in the federal court, including a copy of the permit and the monitoring reports along with a motion for summary judgement, gave a good chance of success. The EPA filed 81 such cases under the Act in 1979. That came down to 32 in 1981, and 14 by 1982.[16] Meanwhile, Jeffrey G. Miller noticed how citizen enforcement suits 'mushroomed … largely as a result of the NRDC effort'. Where the EPA's enforcement of the Clean Water Act had been reduced to 14 cases in 1982, citizens' groups delivered 108 notices of intent to sue, and filed 62 cases.[17]

Sixty of those cases were James's. By the time he took on Gwaltney in 1984, he already had quite a track record of success.

Joseph Luter III took a turn in the family slaughterhouse in Smithfield as a teenager. He fancied studying law himself, but changed plans when his father died during his first year of a business administration degree.

By the age of 26, he was president of Luter Packing Company, and by 1969 had sold the business for $20 million. He came back to what was now Smithfield Foods in 1975 and turned an $8 million loss into a profit inside a year. Luter was a hands on CEO. 'He could walk through a hog processing plant and in 30 minutes tell you what was right, what was wrong and how to fix it,' retail analyst Kenneth Gassman recalled. 'One time, Joe reached right in to lift out some meat with his hands to show us how much less fat there was. I was just trying to keep from running out or throwing up with all the blood there was.'[18]

In 1981, Luter figured it was time to start buying up competitors. He approached ITT Corp, who owned Gwaltney, and wrapped up the acquisition in a day. Gwaltney had been chronically violating its permits for faecal coliform and chlorine discharges into the Pagan River, and would continue to do so for a year before new machinery was installed. From November 1981 to July 1982 those faecal coliform excesses (the reason the state had banned human consumption of shellfish from the Pagan River) were continuous and exceeded the permitted monthly average by 58–150 per cent.[19]

James analysed the company's self-reporting, and in February 1984 notified Gwaltney, the EPA, and the Virginia State Water Control Board of his intent to bring a citizen suit. Together with the Chesapeake Bay Foundation, NRDC filed suit in the US District Court for the Eastern District of Virginia on June 15, 1984.[20]

It was time for James to head to the Gwaltney corporate office in Smithfield, Virginia, and start negotiations.

The temperature was touching 40°C. James's rental car was shadowed by truck after truck, each filled with pigs. The stench permeated the air cooling system.

The public highway turned private, and James drove through Gwaltney's gates. Though his instinct was to hurry from the parking lot to the reception, darting out of the searing heat, James was momentarily stunned on opening his car door. The air was so thick with stench it hit his stomach, and he had to pause to let the nausea pass.

James controlled his reaction to the sheer stink of it all. This visit was not about air quality, nor the ethics of the meat industry. Gwaltney had filed accounts of their own violations, secure in the knowledge that the government would bring no actions against them, but freedom of information meant that these reports were available to citizens on request. James was confident as he headed to the main offices.

The waiting room was well stocked with a full range of magazines related to killing meat for food, including the latest edition of *Industrial Meat Magazine*. James was not kept long. There was no need to be impolite before showing the young pup the door.

Two men waited in the vast office. The larger of the two was Joseph Luter III. He was lodged in a chair behind a desk scattered with little statues of pigs, all getting up to playful antics. A wooden block gave out his name in bold block letters: BIG HAM. To his side, in a rocking chair, was the local attorney Woodrow Crook.

Mr Luter took a while to study his opponent. Young, he noted, thin, all done up in a fancy Paul Stuart seersucker suit with its light blue and white stripes. It shouldn't take long to be rid of him.

'Well, boy,' he said, lingering on 'boy', to show the term was well considered. 'What can I do for you?'

'Well, sir, it's a pleasure to meet you,' James began. And then he stated his case. He had established that Gwaltney had been routinely violating federal law for the past five years, which was as far back as the statute of limitations would allow. Those violations happened on a monthly basis. Having won many cases already, he had no doubt that he would win this case against Gwaltney if forced in that direction. He would then require the company to pay the full price of cleaning up the river, pay up to $5 million in penalties to the US Treasury, and also pay his full attorney's fees, which would be calculated at the rate of a Wall Street lawyer.

Mr Luter blinked and puffed out his cheeks. He made the slow turn towards his lawyer.

'Well, Woody?' he asked. 'What do we do?'

Woodrow Crook had watched the whole play, and summed it up in a sentence as pithy as any Hollywood movie could ask for. 'Well, you can pay him now or you can pay him later, but if you pay him later it's gonna cost you a lot more money.'

Joseph Luter had wanted to go to law school. He was apparently determined to have his day in court, whatever the cost.

Judge Merhige was set to try the case. As a northerner, southern constituents regarded him with suspicion. Gwaltney was represented by Woodrow Crook and a team of lawyers from Richmond, headed by a former Attorney General of the State of Virginia. James was accompanied by Bud Watson, lawyer for the Chesapeake Bay Foundation, the local co-plaintiff in the case.

The two sides were called to the Judge's chambers before proceedings began. Merhige had studied the papers, including Gwaltney's own detailed reports of their emissions violations, and turned to the Gwaltney team. He was happy to put his views on record: 'I have no sympathy for people who pollute.'[21] Merhige announced himself as a Federal Judge in the State of Virginia. As such, Gwaltney were polluting his waterways. They could not do that with impunity. What were they going to do about it?

The plaintiffs had already won some points to establish the right to be in court at all. One of which was *standing*: the right of access to the court system in defence of your own or the public interest. The establishment of such citizen rights is crucial to the furtherance of public interest law as described by this book, and it was crucial to this case. Affidavits from both the Chesapeake Bay Foundation and NRDC established that their members' interests were harmed by pollution of the Pagan River. These formed part of the summary judgement before the trial, which Gwaltney's team made no response to in the two months preceding trial.

The Clean Water Act was just 12 years old at the time of trial. Much environmental law is young in such a way, therefore tender, and is weathered by debate that either enfeebles it or makes it more robust. However tight the language, a law will be interrogated to discover any fault line it contains, with one team of skilled attorneys lined up against another. A point of great contention in this case was the tense in which the Act was written. The citizens' suit provision allowed that 'any citizen may commence a civil action on his own behalf against any person ... who is alleged to be in violation of an effluent standard or limitation under this chapter'.[22]

Gwaltney contended that the present tense, 'to be in violation', was profoundly different to 'to have violated'. They claimed citizens had no standing to bring to court violations that were historic.

Civilisation has grown to depend on the law to the extent that the fate of whole ecosystems now hangs on the use of such linguistic niceties. Without talented lawyers' intense scrutiny of legal language on the Earth's behalf, ecosystems will continue to vanish. James took out the large green tomes of legislative history which contained the Congressional record of debate when the Clean Water Act was passed. The court duly considered the comments he found there from Senator Muskie, who had been the manager of the citizens' suit provision. Muskie stated, 'a citizen has a right under Section 505 to bring an action for an appropriate remedy in the case of any person who is alleged to be, or to have been, in violation, whether the violation be a continuous one, or an occasional or sporadic one'.[23]

The court was satisfied to continue and examine the extent to which Gwaltney's excess pollution was historic or likely to be ongoing.

Gwaltney's own expert witness, a Mr Sneed, entered the witness box to face enquiry from James Thornton. The violations under dispute between the 1981 purchase of Gwaltney and August 30, 1984 included 87 for excess discharges of Total Kjeldahl Nitrogen (TKN),[24] which spurred algal growth and consequent low levels of oxygen (hypoxia) in the Pagan River and Chesapeake Bay. Applying an adequate covering of grease could limit that nitrogen discharge.

'Given the importance, as you mentioned before, of maintaining adequate grease cover, isn't there some doubt in your mind as to whether the Gwaltney facility would be in compliance with TKN limits this winter?' James challenged the Gwaltney witness.

'Yes, we have.'

'Isn't there some doubt?' James persisted. 'That is all I am asking.'

'I think there is some doubt every year that you would expect the plant to go out of compliance at some time,' Mr Sneed conceded.[25]

It was time to bring Joseph Luter III to the stand. James challenged the CEO on a particular violation that had been ongoing for six months. Why was that?

James recalled Mr Luter's simple answer. 'Well, this piece of equipment we're supposed to clean it up with was broken.'

'For six months?'

'Yes, for six months.'

'Well, sir, let me enquire: what would be the most important piece of equipment to produce the hams that you produce every day in your factory?'

Mr Luter had real expertise in terms of such equipment. He knew a whole range of possible answers, just not which one it was best to give at this point. 'Well, I don't know what the most important one would be.'

'Well, let me suggest some to you. For example, the pig stunner with which you kill the pigs, would that be a very important piece of equipment?'

'Why, yes.'

'Well, if all of your pig stunners went on the blink, would you have waited six months before you repaired them?'

James considered the next moment to be just like a comic book. It felt like being met with a stream of little daggers shooting out from the enemy's eyes. A favourite joke of Luter's has the punchline 'if a man didn't have any enemies, he didn't do a damn thing with his life'.[26] James had just acquired himself an enemy. Joseph Luter III looked daggers at the young pup, and said nothing.

'I'm sorry,' Judge Merhige prompted, 'but you must answer Mr Thornton's question.'

Mr Luter was no fool. He knew this was a case of a fish being asked to bite a big hook, and he was no fish. Still, he was forced to answer. 'No, I would not wait for six months to make them work.'

'Well, do tell me, how long would you wait?' James pursued. 'Would you wait five months or three months or two months? Would you wait two days or would you have them fixed that day?'

'Well, I'd have to have them fixed that day.'

'Then what's the difference between one day and six months if this was important to you actually complying with the law? Was complying with the law not important to you?'

'Objection.'

Objections were overruled.

The arguments then proceeded to establish appropriate penalties, which are a core part of the legal strategy that this book will proceed to outline. 'Corporations speak in the grammar of money,' is James's basic statement of the tenet. 'It's their only grammar. Since they are very happy to violate the environmental laws, if you want them to take those laws seriously you need to speak to them in the grammar of money, and make them pay a great deal of money for violating the environmental laws. Then suddenly they'll wake up to it.'

James's question of Joseph Luter in court, as to how long he might take to repair a piece of equipment he deemed vital to a plant's operation, proved resonant in the Judge's summary. Contrary to Gwaltney's contention, Merhige concluded:

The Court believes that Gwaltney's penalty ought to be increased, not reduced, because of wilfulness. Gwaltney's lackadaisical approach in correcting a problem that posed risks — albeit not 'imminent' ones — to both human health and aquatic life should not be countenanced. One may speculate how long Gwaltney would have taken to repair a

machine the faulty operation of which would have halted production
… [A]t the very least Gwaltney would have exerted more effort to
repair such a machine than it did to bring its discharge into compliance
with pollution standards.[27]

Based on 666 days of violation, an award was made against Gwaltney
for a total penalty of $1,285,322, 'most of which was designed to deter
possible future violations by Gwaltney or other similarly situated
polluters'.[28]

A 'Save the Bay' campaign was announced in December 1983, and
received considerable state and federal funding. Its main focus was the
run off of agricultural pesticides and fertilisers. The public noticed the
absence of dirty foam and floating debris and presumed the bay was
returning to good health. However, one difficulty with campaigns for
raising public awareness in the late 20th and early 21st centuries is that
much of the toxic pollution is invisible. Poisons dumped in the bay had
accumulated there, and it was dying.[29]

Bethlehem Steel had their iron and steel plant on 930 hectares of
land in Sparrows Point, Maryland. They discharged one billion litres
of wastewater into the bay every day — the equivalent of a small river.
When permits were first introduced, industries were allowed to negotiate
their own limits based on the use of the best available technology, and
what was affordable rather than what was beneficial for the water
quality.[30] Bethlehem Steel was therefore permitted to dump 18 kilograms
of cyanide a day into the bay, alongside 32,000 tonnes of oil and grease,
and hundreds of kilograms of chromium, iron, and zinc.[31] Industries in
Maryland and Virginia dumped 2,891 tonnes of heavy metals into the
bay every year. Half of the toxics that Baltimore industries discharged
into the bay were credited to Bethlehem Steel. Scientists who tested the
mud at the bottom of Baltimore Harbor found it contained 480 different
toxic chemicals.[32]

James filed suit against Bethlehem Steel in April 1984, while still in the middle of the case against Gwaltney and an array of other suits. The local co-plaintiff was again the Chesapeake Bay Foundation, this time with Scott Burns on the case.

The complaint detailed continuous violations of Bethlehem's permit since 1979. When the case moved to discovery, the pre-trial procedure through which opposing parties in a lawsuit seek to obtain evidence from each other, James went to the Bethlehem Steel headquarters and was shown into two large rooms stacked high with documentation. One favourite tactic in litigation is to overwhelm the opposing team's legal resources. Instead of feeling overwhelmed by the weight of material, James ordered thousands of pages of photocopies. These filled two rooms of his New York base, and he examined them throughout his evenings and weekends.

His complaint to the court was then amended to account for the hundreds more violations that were now discovered. The documents also revealed that the company used an instrument they knew to be faulty in order to underestimate discharges of chemicals and metals in what amounted to 'fraudulent concealment'.

James brought to the court's attention another dark gem from his research: an internal memo from a Bethlehem Steel official that recommended against spending up to $400,000 a year to meet permit requirements for a treatment plant. 'Just the thought of expending funds to meet an unrealistic limit that should have been changed years ago is grotesque,' the official complained.[33] Scrutiny of the company's documents meant that the severity of the violations was supported by over 100,000 pieces of evidence.[34]

The summary judgement was upheld, and the case moved on to trial to determine the level of penalties. Contending that Bethlehem Steel had purposefully polluted the bay to save itself up to $36 million, James demanded the company pay $160 million in penalties to the US Treasury. The company had a reported income of $34 million for the last quarter, but an annual loss of $153 million.

Fellow polluters were able to buy into a $5,000 a year industry newsletter that tracked the progress of James's work with the Citizens' Enforcement Project in order to learn whether it made better economic sense to clean up their activities and stop permit violations, or risk being dragged into the courts. A sizeable penalty in this instance would put a brake not only on Bethlehem's serial violations of their permits, but affect the actions of others who were watching the outcome.

Lawyers from a top Baltimore firm negotiated hard up to the very eve of the trial. They settled for a payment in lieu of penalty to Chesapeake environmental charities of $1 million, then the largest ever settlement to a third party environmental fund.[35] In addition, they paid $500,000 in attorneys' fees, to cover James's hourly rates pegged to top Maryland lawyer rates. It was noted how 'other industrial polluters began clean-ups to avoid similar suits'.[36]

On the day of the settlement, James had a private meeting with a white haired Senior Vice President of Bethlehem Steel, who had reached an age that saw him set for retirement. 'Young man,' he said, 'we just never thought that we would be having this meeting with you in which we are settling for large sums of money, because we had a deal. A deal with the federal government in which we knew they would not prosecute us, and we knew the State of Maryland would not prosecute us. It had never occurred to us that a citizen like yourself could show up and haul us before the court and win. You have really taught us something.'

Michael McIntosh's aim was to loan James's start-up money for the programme so that it would open up a separate funding stream for the parent organisation. James kept a separate bank account for the Citizens' Enforcement Project. In addition to the millions paid directly to environmental charities in lieu of penalties, the award of attorneys' fees meant that all wages, costs, and expenses were paid, and the loan was repaid to the McIntosh Foundation. The project was still in profit, and James, working for charity wages, wrote a cheque for $1 million and handed it to NRDC.

However, James had two additional goals. One was to establish citizen suits as a viable alternative form of environmental enforcement and the other was to goad the government regulator, the EPA, back into doing its job.

'I wanted to win my cases,' James recalled. 'That was a lot of fun and I did win my cases, but the main ambition really was to use the winning of my cases as leverage to force the government to start enforcing the law again. That was my strategy. The government has ten million times the power of one lawyer to make the laws work, but if unchecked, the government will always drift towards what the companies want, because companies are fantastically more powerful than citizens.'

Jeffrey G. Miller has written how the EPA conducted an internal review of James's work for NRDC, because 'if a private citizens' group could commence as many enforcement actions as quickly as NRDC and its allies had, the question naturally arose whether EPA had fallen down on its job or whether NRDC was simply pursuing minor problems'. The study 'concluded that 10–15 percent of the files reviewed by NRDC resulted in notices of violation and that a "significant number" of notices were followed by the filing of a complaint. It found that many of the targeted permitters had what EPA considered to be significant violations.'[37]

In 1983, press coverage of the dire situation at the Environmental Protection Agency saw Anne Gorsuch forced to leave her post and the founding administrator, William Ruckelshaus, return. He was horrified by what he found his predecessor had done, remarking, 'I mean, it really was awful. If anything, the press underplayed its seriousness.'[38]

One Agency staffer noted at the time, 'Our biggest problem right now is that none of the old compliance people are left. They all got moved out.' Ruckelshaus went on to address his enforcement staff in 1984. 'We can find 100 reasons not to do something in terms of organizational structure, guidance, you name it,' he told them. 'There ought to be 100 reasons to do something. We have to develop a certain controlled sense of outrage in this agency if we are going to get these laws enforced. And some place along the way, we have lost that.'[39]

James was called to Washington to meet with Ruckelshaus. He presumed the meeting meant a quiet sandwich in Ruckelshaus's office, but was instead led into a conference room filled with upwards of 150 EPA enforcement lawyers. There were sandwiches for all. Ruckelshaus introduced James to the group and then said, 'Mr Thornton, now could you give us a seminar on how to do enforcement cases? You have kept the torch burning while it has gone out here, and we would wish to learn from your experience how to bring and successfully prosecute good cases. Over to you.'

And so James gave the government lawyers an hour and a half seminar on how to win their cases, which felt like a sign he had accomplished one of his project's central purposes.

James developed legal cases against 88 major violators, including such corporate Goliaths as Pfizer, Upjohn, General Electric, and Texas Instruments. That meant investigating the pollution of waters by more than 1,000 companies. In Michigan, he alerted the state to his intent to file suits against 20 major polluters, including the big three auto companies. The state accepted responsibility for the cases and imposed $1.86 million in penalties on the firms. In sum total, 44 cases were settled out of court, a further 30 were won by James for NRDC following trial, and 14 were passed on to others to complete when James chose to move on, job done.[40]

Each case was part of an unfolding strategy that paved the way for the next. The trial against Bethlehem Steel, for instance, established that the 'EPA looks to citizen suits to supplement enforcement because the EPA and state agencies lack sufficient resources to bring all necessary actions'.

Joseph Luter III took his case all the way to the Supreme Court, where the judges gave some ground to his lawyers' arguments on use of the present tense. However, Justice Marshall allowed that '"the practical difficulties of detecting and proving chronic episodic violations of environmental standards" suggested that "a good-faith allegation of ongoing violation" sufficed to establish jurisdiction over a citizen suit'.[41] The case against Gwaltney gave such strong proof of continuing violations that penalties stood, to be reassessed and then confirmed back in the District Court. After trials and hearings in the District Court, the Court of Appeal, the Supreme Court, the Court of Appeal again, and then back down to the District Court, Gwaltney's guilt and penalty were confirmed.

In his *Rolling Stone* article, the reporter Jeff Tietz asserted that 'ostentatious pollution is a linchpin of Smithfield's business model', the argument being that volumes of waste from hog production are so vast and so toxic that no environmentally sound waste treatment could be economically viable. Joseph Luter's Smithfield company disputed this, but continued to discharge excess pollution. The question was whether the Environmental Protection Agency was now effectively forced back into its role as government regulator. 'The bay really is as everyone's been saying … a national treasure,' Ruckelshaus allowed in 1986. 'But it's also a national challenge. As of this moment, it's an open question as to whether we're going to rise to meet that challenge.'[42]

In 1996, almost ten years after James's Supreme Court victory against Gwaltney, the company's continuing bad actions meant that the federal government filed a $125 million water pollution lawsuit against it. At a point where the EPA had documented 75 violations by Smithfield, Joseph Luter calculated that the number of violations Smithfield could theoretically have been charged with was 2.5 million. 'A very, very small percent' of the actionable violations were raised by the EPA, he boasted.[43] Luter called a press conference in response to the suit at which he spoke for an hour. It was proposed that he settle the suit for $3.5 million. Of course, he once again rode the appeals route all the way to the Supreme Court. His company was obliged to pay a $12.6 million judgement.[44]

The relevance of those early cases never faded for James. 'Even well intentioned governments have enormous pressure from the other side, the other side being in most cases industry. There are some good companies, but let's be honest: most industries, particularly most polluting industries, don't want to spend a penny on compliance. Often, there aren't laws, but where there are laws they would much prefer to yell at the government to not enforce them. If you are going to make the laws that exist work, you really need citizens to have the ability to enforce the laws when the government doesn't do so. Without that, there isn't a constituency. These are very complex issues, and citizens don't have any information. They are not going to demand the government enforce the laws, because they won't even know the laws are being violated.'

This is what sets legal environmental law groups apart from environmental campaigning organisations. 'They have the expertise to analyse the science, understand the law, and actually force implementation and enforcement where it isn't being done.'

Pollution indicators in Chesapeake Bay, checking on levels of nitrogen, phosphorus, dissolved oxygen, water clarity, and toxics, showed an 11 per cent year on year improvement in 2014. The Chesapeake Bay Foundation now runs a 'State of the Bay' report. The target figure of 100 represents the bay in the 1600s, when Captain John Smith sailed in with those early explorers. Oyster reefs, composed of creatures who are able to filter the bay back to purity in under four days, are unlikely to build high once again, but it's a good benchmark to have. In 1999, the bay scored 28, but by 2014 it had crept up to 32, leaving 68 points to go.[45]

Stop a company dumping toxics into a river and you gain downstream benefits. The ecosystem of a whole lake or vast bay might well be enhanced. The Citizens' Enforcement Project did trigger such benefits through a piecemeal approach: it identified the major polluters and stopped them one by one until the government enforcer resumed its proper role.

It takes huge resources to sustain enforcement on a case by case model. An alternative is an ecosystems based approach. James left New York and headed to the West Coast, where such an opportunity awaited.

We are made of law

James Thornton

Law and people go together like tail wagging and dogs. It is not that using law shows we are happy, but rather that you can't separate the two.

Tracing the development, some ten thousand years ago, of small scale hunter gathering into farming, reveals the early existence of law. Whenever you have two or more people, there are disputes. Ways to resolve those disputes evolved, and law appeared among us. As we settled in towns and farming produced surpluses of food, we had the luxury of being able to create more complex roles. We differentiated into kings and nobles, priests and administrators. Laws became elaborate and grew more explicit as our culture developed. Hammurabi is recalled as the first lawgiver of this more sophisticated time, and his laws were clear, if severe. Nowadays, it is rare to punish offenders with impalement.

By striving for self-expression, wealth, progeny, and security, we have built civilisation in all the variety we find it on our planet today. The diversity of our cultures enriches the biodiversity that nature has been building for some three and a half billion years. Nature's book of life has encoded within it laws that are older, subtler, and more complex than our own, and it is the project of science to probe these laws that allow matter to do certain things but not others.

In our realm, law is the mutual agreement to bind our actions, and like the laws of science it allows us to do certain things but not others. It channels the force of human enterprise in directions we agree are in our mutual interest to pursue.

Until about 45 years ago, no one saw a need for comprehensive laws about our interaction with the rest of nature. Throughout our evolution, our impacts had been slight, with our efforts directed towards the advancement of our interests. Altruistic efforts largely focused on alleviating poverty and injustice.

With the recognition that our enterprise was eroding the commons, all that changed. We depend on the common resources of clean air, water, wild fish in the sea — and all the rest. One day we woke up and realised these were under threat, and the response was a creative burst of lawmaking. In the 1970s, comprehensive laws were enacted to protect the environment in the United States, and in the 1980s Europe followed.

Those of us who struggle to implement, enforce, and improve these laws were heartened. We had a mission. We had tools. Our opponents were the ignorance and greed that resides in all our hearts, and the resident myopia of politics. But given time, we thought, we will reach the right balance.

The science kept accumulating. The diversity of life was under threat in a way more profound than we had realised, and what was worse — the world was warming. By today, everyone with their eyes open — including the Chinese Politburo — recognises that civilisation hangs in the balance. If the world warms by an increment that we are now on track for, say 4°C, civilisation as we know it is likely to suffer a rolling collapse. Environmental work now is to save civilisation and as many species as we can carry with us.

Law becomes about saving civilisation. It offers the opportunity to delay the impacts of our past actions, and release our creativity to build a positive future together. That is my daily practice. It is also fun.

Law is the answer to the question I'm often asked: what can I do about global problems? Climate change is so big, people say, it leads to despair. Why not just pull down the blinds and watch TV?

The answer is that we have the whole world to fight for. My client is the Earth, and all who live upon her. Nothing makes me happier than standing in a tropical forest and being surrounded by a mixed flock of birds, most of them new to me, each pursuing its separate living. Or perhaps being in a skiff in the waters off Alaska, the water boiling with krill and fish, and a humpback languorously feeding. Or watching a child encounter a squirrel for the first time in an urban park.

My client has specific needs. She needs us to reduce our carbon emissions and clean up our pollution, to reduce our impacts on the diversity of life. We need to save the bees and teach people that our salvation depends on intimately taking care of all life.

Isn't this a wonderful practice? This is law to me. It starts with the science, which is the grammar in which the Earth speaks to us. To understand the needs of my client, I talk to scientists. Environmental law tries to capture, in a snapshot, the best science that can be turned into policy, and then into rule of law. Next, we need to implement these laws and enforce them.

I have been fortunate to work for the rule of law. Cases I have brought and caused to be brought have established the right of citizens to protect the environment in the Supreme Courts of the United States, Poland, and the United Kingdom, as well as the highest court of the European Union. A recent delight has been advising the Chinese Supreme People's Court on the interpretation of their new Environmental Protection Law, which will allow citizens to sue polluters. As far as I know, no one else has had this range of privilege.

It is citizens that make environmental law work. Their desires create the political will to allow the laws to be passed, and they need to be in the legislature while the laws are being written. When the laws are on the books, citizens need to implement and enforce them. Inevitably, economic interests press from the other side, and even well intentioned governments fall short unless citizens push to make the laws work.

The forces to do this work are small. There are not many lawyers who take the Earth as their client. Strategy, therefore, is crucial. With such small forces, you must be highly strategic, as any good general knows.

My client is in need of help, and much can be done. We are always looking ahead to what the law needs to be in order to keep the planet safe for our children and other wild things. Environmental law is less than half a century old, and so it is incomplete. Even if all existing environmental law was properly enforced, it would not stop climate change and the loss of species, only reduce it.

Therefore, we work in parallel process: making the current laws work to create a real rule of law for the global environment, and improving the existing laws, moving our collective behaviour to eradicate actions that damage the fabric of civilisation.

I take hope. I find it in the practice of environmental law, because of the gains made every day. I find hope too in that so many eyes are opening. In the 1970s, it was the United States; in the 1980s, Europe. Now it is China. The Chinese see protection of the environment as central to the future of their civilisation. They are right, and I believe that they will move heaven and Earth to take care of it.

Environmental problems are human problems and they arise from our conduct. This means we can solve them by changing our views. We do not need enlightenment to do this, only enlightened self-interest, and it must speak louder to an ever widening circle to lead the ambition for change in the right direction.

Only law can capture that ambition, make it systematic and enforceable. I want a rule of law on Earth that leads as ineluctably to the protection of civilisation and our companion species as Newton's laws govern motion. Will you join this endeavour? My client tells me that she wants it too.

2

Where the Wild Things Are

Malibu Beach real estate looks unassuming from the road. Garages and front doors open out onto the pavement, and the constant flow of traffic stains the houses' low outer walls. Step through a front door, though, and light reflects off marble floors. You'll likely pass an Oscar statuette, for this is Hollywood mogul land, on your way to the view. This is what many millions of dollars can buy you: your own plate glass panorama of the Pacific Ocean.

James knew the area from knocking on Hollywood's front doors. The producers Sydney Pollack, Frank Wells, and Alan Horn were among his early supporters, as James pulled in funds to start a Los Angeles office of NRDC. Barbra Streisand had one of the finest Pacific views. Meryl Streep gave him her private bedside phone number. I enjoyed all the stories of the stars.

Up to the right of us was Corral Canyon. Bob Hope was the star involved in that one. He felt it should be developed into a golf course. James liked to quote the line in the Environmental Impact Statement the comedian submitted along with his proposal. 'This meets the unmet need for a world-class championship golf course in Malibu.' Like many of Hope's lines, that one is worth a laugh. The golf course plan was defeated, and Corral Canyon is now a park.

This was 1994 though, and that fight was in the past. We were speeding north in James's old black Honda Accord. We were on his escape route. The northern reaches of Malibu relax into larger, more verdant estates, and then development ceases. The Pacific breaks against giant boulders to the left of the road, and we turned right. The old car crunched through a dirt car park.

James pulled open the boot of the car and took out his walking boots. He couldn't stop grinning. So this was it — the entrance to one of his favourite walks in the world. He came here for snatches of nature when urban Los Angeles threatened to swallow him whole. It was grand to be back.

I'm English. I don't enthuse lightly. I reserved judgement, and followed James along the dirt trail into Sycamore Canyon Wilderness Park.

Earthquakes carved out this canyon and water once thundered down it, but now the trail follows a thin stream. Large sycamores grow among willow and cottonwood at the edges, and hikers pause for photos beside their trunks. Yuccas sprout six metre high blooms on the hillside. 'Breathe that in,' James said, and so I did. It was the herbal smell of sage among the sagebrush.[1]

I was beginning to get it. We turned right and climbed to leave the family groups behind. The landscape opened out into the rolling grassland of California's hills. Above, where the path curved left, I spotted movement. Was it a dog, a large dog? It paused and raised its giant ears. This was my first coyote. It fixed me with its eyes. For a minute or two, we stood and stared. This was wilderness, and the coyote taught me what the word meant. I was staring into the wild.

The coyote broke the gaze, shook its head, and loped off. We continued up the trail. A bird was poised above the grassy peak of the summit, a white tailed kite, its broad white wings stretched as it hovered.

The loop of our trail took us high above the road, with views between large rocks to the ocean. A quick dart of blue-grey, and we focused on a creature that hopped between the rocks and branches beside our path. This was a California gnatcatcher, a rare visitor to this canyon. I knew of this species. It first glimmered into view in James's tales of how he had saved thousands of hectares of coastal California. That was in tracts of

Orange County, on the southern side of Los Angeles. Malibu Beach and its surrounding canyons showed developers how to turn the wilderness into multi-million dollar profits. They had done all they could to the north. Southern expansion was the way to go.

This California gnatcatcher was moving company as we ended my first walk in California wilderness. It opened a rare tale, of how man and bird learned to share habitat.

James had moved out to the West Coast to be in the same time zone as his teacher, the Japanese Zen master Maezumi Roshi. I met Maezumi on a California mountain in that summer of 1994, and thanked him for his teachings to James. 'I am not his teacher,' Maezumi replied, and swept his hands about him to include the tree clad mountainsides. 'Nature. Nature is his teacher.'

James worked from NRDC's San Francisco office. Other major groups had tried to set up an environmental law office in LA, and failed. 'Move closer,' Maezumi challenged James on one of his visits. 'Set up an office in Los Angeles.'

The rules were set. NRDC allowed James one year on salary to do the weekly commute from San Francisco and see what he could bring off, but he must not approach their current funders. All funding had to be new. He met with one potential funder in her Malibu home. 'Will you be like the others, ask for our money, and then leave? Or will you stay?' she asked.

'Stay,' he said, and borrowed a friend's sofa before moving into a studio in the LA Zen Centre.

Leap on a year or two, and James had his own NRDC office in the art deco Oviatt Building in downtown Los Angeles. The office had its start-up strategy in place. A small staff worked on such issues as air and water quality, lead poisoning of inner city children, and transportation. It was time for James to scale up his meetings.

He drove east. The air started to clear. The Sierras ranged to his left. Beethoven's violin concerto spilled out of his car radio. The world felt expansive. He raced along the freeway and took the Pasadena exit.

James waited in the lot of the California Institute of Technology for the violin concerto to finish. It stirred him with a sense of purpose. Caltech is one of the world's very top universities. The physicist Richard Feynman set the standard when he moved there after playing his part in the wartime Manhattan Project in Los Alamos. In 1955, the physicist Murray Gell-Mann took his first tenured post at Caltech so as to work alongside Feynman. Gell-Mann was awarded the 1969 Nobel Prize for Physics for his achievement in postulating the existence of quarks. A guy who could determine the existence of something no one has ever laid eyes on could well guide James's new project in some helpful way.

Just touching 60, Gell-Mann was swarthy, his head crowned by a foam of white hair, and his eyes intense behind thick black rimmed glasses. He was on the board of the MacArthur Foundation, which had global conservation as one of its aims, and so that gave the two men a shared platform on which to build. James opened on what he thought to be safe ground, and spoke of his passion for protecting tropical rainforests.

'Why does everyone talk about tropical rainforests?' Gell-Mann exclaimed. He had retained his New York Jewish accent. 'Why don't you take care of tropical dry forests — they're almost as species diverse.'

Gell-Mann had his theme. James could try and vary the tack all he liked, without much sense of being listened to. Gell-Mann returned to the need to protect biodiversity, like he had found the stick with which he could beat the younger man.

Gell-Mann's early wish to partner with Feynman had soon developed into rivalry, which morphed into 'a bitter feud'. Gell-Mann became 'notorious for his bad temper. Among his cronies, the nasty names he coined for rivals were as familiar as the catchy terminology he applied to particles.'[2] It's hard to lay a lifetime's commitment to rivalry to rest. James sensed his lunchtime had become a competition, with the scientist out to prove to the lawyer who was the better environmentalist.

Job done, Gell-Mann used lunch to regale James with tales of the Korean practice of slapping naked female buttocks with wet silk. The meeting ended. The men parted.

In his car on the way back to LA, James steamed with annoyance. Tropical dry forests were one thing, but where was he going to find equivalent biodiversity in LA? Gell-Mann had barely listened to him, but just kept repeating his challenge. 'Some of one's best creative efforts come from getting annoyed,' James would later allow.

Gell-Mann's challenge bit hard on James's psyche. His favourite walk in the area took him through Sycamore Canyon, a protected wilderness beyond Malibu where Los Angeles' northward expansion finally takes a break. The habitat changes as it rises from the coast and into the hills. This is coastal scrub, a Californian equivalent of what the French call garrigue — it's not tropical dry forest, but it's pretty dry and has its own rich biodiversity.

South of here, the scrub vanished beneath Los Angeles' urban sprawl, but for 68 kilometres south of Long Beach the density of building relaxed. The coastline of Orange County contained six small 'surf cities', and between them were Southern California's last undeveloped canyons. Oak and sycamore gathered into woodlands. Flowers decorated the sage scrub that covered hilltops and slopes. Mountain lions fed off deer, rattlesnakes coiled up in native valley grassland, and coyotes howled into the night. This was a great place for mountain chaparral, a community of hard leafed plants that prosper in semi-arid zones.

Droughts in California are cyclical. One blasted the area in the 1860s, which gave a buying opportunity to James Irvine and his partners. They purchased vast tracts of land and ran them as ranches for more than a century. Life changed, Hollywood happened, beach life became ever more desirable, and property values soared. The Irvine Company sold its land holdings to a developer in 1983.

This whole natural world was now parcelled out on architectural drawing boards. Los Angeles was looking to grow, and developers had the ownership rights. Urbanisation was set to roll back from the shore till it wiped out all wilderness from LA to San Diego.

James is hugely competitive, one of the standard aspects of entrepreneurs. That's why Gell-Mann's jousting took such effect. However, James couples competitiveness with training in a more reflective mode. Taking a good walk helps.

It wasn't feasible to block this mass development of the coastal zone lot by lot, issue by issue. James had the freedom to direct some of his own time, but had no other resource in place. He walked out into the chaparral. This was completely beautiful, undeveloped, natural and wild, while just a kilometre or so away dense housing crowded in and destroyed 100 per cent of the ecological value.

If James did not take some action, the whole California coast would be lost to development. 'Instead of thinking through policy options and political realities, I started to meditate on it,' James later wrote. 'As I visited the undeveloped land in my mind, I noticed a feature I had never thought about before. There was a very clear coastal zone that went from the ocean to the first ridge of hills.'[3]

This was in fact a shift away from a simple focus on a problem, which at its simplest was the threat of a bulldozer tearing through the land, an army of bulldozers each with its own attack plan. Appreciation of the natural world replaced the urgent problem of habitat destruction for a while. This is what a good lawyer does — takes time with the client so as to gather the wider picture rather than simply sharing the client's sense of threat.

Now the understanding of a coastal zone was in place, James could see a possible solution. Surely some bird or mammal, ideally one with a touch of charisma so it could harness popular support, had its home uniquely here.[4] Loss of its habitat would then threaten its existence. James could then petition to have the creature enlisted as an endangered species, and so unleash federal protections.

He read and he studied and he asked around. No creature emerged. He needed to add a separate research technique.

James's home in LA was a small condo attached to the LA Zen Centre in Korea Town. Night-times were cracked open by gunfire as drug wars were waged on the streets outside. Helicopters churned overhead and beamed searchlights down onto the streets and backyards. It took discipline to go quiet in a cityscape as loud and violent as that one. James trained with Maezumi Roshi, and meditation became a serious practice.

One aspect of the Zen tradition is koan practice: you settle a question in your mind and sit, hour after hour, to see what answer emerges. James settled with the question of how to preserve the coastline of Southern California. After one particularly long session, he rose from his cushion and went to the phone. Meditation had suggested a plan of action, and he went straight to it. His call went through to the Cornell Laboratory of Ornithology. Did they know of a bird whose life was coterminous with the California coastal zone?

Not only did they know of one, a colleague had just completed a monograph that identified this bird as a subspecies endemic to the region.[5] The California gnatcatcher is a ten centimetre long dusky grey songbird, its song being one of its most defining features. It sounds like the mewing of a kitten. The birds mate for life and never migrate. An estimated 2,600 pairs made their life within southern California's coastal scrub.

James gathered in the science, and with the aid of legal intern Mike Thrift wrote the petition to the US Fish and Wildlife Service to list the bird as threatened. Within three weeks of its submission, he was in meetings with senior officers of the Irvine Company at their headquarters. He explained how he had the power to get the bird listed as endangered and thereby shut down all development. Instead, he was prepared to work out an amicable deal in which they might proceed with some managed development while the biodiversity was saved.

Why the compromise?

'I'm a pragmatist,' is James's response. 'It would have been impossible to stop all development forever. You have to harness the energy. With the forces I had, even with the law being good, it would have been seen as being incredibly unreasonable. You would have looked like a complete fool to try and prevent all development of the Californian coast from LA to the Mexican border for any reason, let alone to save one small bird. I wanted to get a result that I thought was achievable and could be maintained in perpetuity, rather than something that you might win for a little while and then would get swept away. Which certainly would have happened. First of all, you would have lost if you'd tried to block all development. Or if it worked for a little while, it would have been seen as a suppression of the needs of Californians for housing, let alone economic development. It would have been chipped away and eventually destroyed.'

The negotiations that followed, led for NRDC by Ann Notthoff, led to California's Natural Communities Conservation Planning (NCCP). This was a process through which 'developers are allowed to destroy habitat and "take" some small number of California Gnatcatchers or other of the 77 Endangered or Threatened plant and animal species that occur in the region'. In exchange, they set aside larger amounts of land for conservation. 'As of 2001, at least 110,000 acres [44,500 hectares] of natural habitat including 72,000 acres [29,000 hectares] of coastal sage scrub had been placed in conservation status under six plans approved through the NCCP process.'[6]

The California gnatcatcher was listed as threatened rather than endangered in 1993, so as to allow the state to consider whole ecosystems rather than just a single species when deciding how to parcel habitat for humans and people. This was seen as a way of repairing use of the Endangered Species Act, which had lost the traction of popular support through a prolonged case to preserve the spotted owl, a dispute that halted lumbering for the timber communities of the Pacific Northwest.

Steve Johnson, a spokesman for the charitable environmental organisation Nature Conservancy, said of this new approach: 'If it serves as a model for the nation, it will provide a much more practical method of solving endangered species problems than we had before. We think it's great.' The US Interior Secretary Bruce Babbitt called this new approach 'a "trailblazing effort" to mesh the Endangered Species Act with economic needs, and said the California effort would "provide some pointers to how we handle future issues."'[7]

One awkward fact which you'll notice as a threnody throughout this book is that no environmental battle is ever won, even one based on the premise of amicable compromise. Economic self-interest will always try to assert itself. In 2010 and again in 2014, the conservative Pacific Legal Foundation, representing two groups of property owners, petitioned to have the California gnatcatcher delisted. This was on the basis of a genetic study they had funded, challenged by other scientists, which found genetic similarities with the gnatcatcher of Baja, Mexico, that belied differences in physical appearance.[8] Their argument is essentially for the extinction of the bird and its habitat in the United States.[9]

James reflected on his approach of taking an issue into meditation after days of wrestling with it in his head. 'I use a creative process which combines a high degree of analytical skill and input with a creative non-cognitive process,' he said. 'For me, I use meditation, though taking a hike in nature can do it as well. You need to go wide in order to see what the problems are and to recognise the field of opportunity that offers a creative solution. That process then allows me to explain what I am doing to non-lawyers. My way of solving problems is different from most lawyers.'

That field of opportunity, of course, appears in societies where the rule of law exists. 'Law is a tool that I use to answer the bigger questions. You need a fairly well ordered society before you can start protecting the environment. You may need to build the rule of law before you can get a solution.'

Public interest law means top lawyers work in our best interests for no fee. Sounds good. What's to lose?

As with everything to do with the environment, it pays to look wider. No lawyer stands alone. No law comes into being without cost. No legal strategy can survive without investment. Serious philanthropists don't give away money. They invest. They apply ruthless business strategy. It's worth a while to consider what drives such money.

James's next shift was to Europe, to ensure that the rule of law was appropriately in place. There he would reconnect with the man who funded his Clean Water work.

Building an ecological civilisation

James Thornton

We need a positive vision.

When you glimpse ecological crisis, the first reaction may be despair and anger. Much of the environmental movement still works against things. Waking up to the problems was the first step. Opposing the problems is the next.

Opposition alone is never enough though. We need a positive vision to pull us towards a future we want to inhabit. I use the word 'vision' consciously, to suggest something beyond the rational. For much of the last half century of activity, the environmental movement has believed that rational argument alone would convince. If we only analysed all problems into their parts and explained our reasoning, we would convince everyone to follow.

Reason remains the foundation. This book describes how we study science and facts to deploy carefully constructed legal arguments and write laws. This work only achieves its greatest power, however, in service to a positive vision. We need a story of where we are going.

An academic lawyer argued to me that 'rule of law' is such a vision. I disagree. Rule of law is the basis for a stable society. But you could create a society with clear rule of law and transparent institutions, yet it could be despotic and environmentally ruinous. Rule of law *serves* a vision.

I built ClientEarth to create positive solutions. We put a lot of effort, for example, into making the world safe for renewable energy. We work on the policy, on the finance, on European and national legislation. We work to stop investment from getting locked into the wrong infrastructure. So we helped stop a new generation of coal fired power plants in the UK, and are doing the same in Poland. Halting bad carbon investments gives society the time to realise the need for renewables, and allows investment the opportunity to flow towards them. Stopping bad decisions can be a necessary precondition for creating a positive solution.

How do you drive enough positive solutions quickly enough? Here is where the vision comes in. A culture has a story. It is the story that determines what we think about, where we look to focus our energies. If the story is the right one, it can liberate boundless energy. It is that motivating story we need, one which building the rule of law can serve, one that allows an endless diversity of positive solutions to emerge, because they become what people think of naturally.

A story of this kind directs our attention for the long term. Consider three levels of temporal scale. On the short term of a few years, we are concerned with the next election, what politician is in power, how much a company's earnings are going up or down, where the euro and renminbi are against the dollar.

On the longer scale of several decades, we see larger patterns begin to cohere. So we had the New Deal in the US after the Great Depression. We had the exaltation of greed in the Reagan and Thatcher dominated decades. We had the experiment of Communism in Eastern Europe after World War II.

The story I am talking about operates on a longer time scale yet. Ten thousand years ago we invented a new way of life when we settled down to agriculture. Towns then cities were born. Religions grew up that fostered centralisation. We looked up towards central authorities in the state and in the afterlife. There was a powerful set of stories around the world, all with the same elements, that made sense of our life. We toiled, built the city, and looked forward to paradise.

In a great shift of story, we moved from agriculture to industry. We replaced the religions that made us look to the mountain with economics that suggested we look to our wallets. A vision of working towards a greater good was replaced by the current vision of our globalised culture. The meaning of life is material gain. Your value is measured by your investments. What is not monetised does not have value. All living things are resources that should be converted to capital, the sooner the better.

Our current story has served us well in advancing materially. It has not served us well in taking care of the deeply complex web of life on our planet. We are only beginning to understand life in its fragility, resilience, and boundless evolutionary creativity. So we need to edit our story.

If the agricultural revolution gave us a story rich with religions and institutions, and the industrial revolution gave us a story rich with economics and material progress, where is the next story to come from? It will be born from our entrepreneurship. Human beings are entrepreneurs. Our story in any epoch is the one that guides and grows out of where our entrepreneurial flair is taking us.

I have been searching for years for the title of our new story. I believe I have found it in the Chinese Constitution. Facing the ruin of their environment, the Chinese looked hard and amended their constitution. This core document now calls for the building of an Ecological Civilisation. This is the best formulation I have heard of what the new story should be. If we are going to thrive into the future, it will be as an ecological civilisation. We built an agricultural, then an industrial, and now must build an ecological civilisation.

Systematically, the Chinese are considering its elements. I sit on an expert panel advising the Chinese Premier on what rule of law for ecological civilisation should look like.

We need to do many things quickly. A new story, as we start to tell it, will guide all our complexly divergent activities into the right channels. A new story gives us the chance we need. If you can come up with a better story than ecological civilisation, I encourage you to share it. If not, start telling the story of ecological civilisation and let us see where it takes us.

3

Leveraging Alaska

Outside, the Tongass National Forest lines the shore. Inside, the philanthropist Michael McIntosh peers into a screen. He's checking the markets, adjusting his investment portfolio. Wi-fi through the Alaskan archipelago on the Juneau to Sitka route is a crapshoot. You grab a signal when you can.

We guests start wondering where he's got to. Twenty of us are served by a crew of 13, including a naturalist who helps direct our binoculars. Those white specks high on the hills? They're mountain goats. Those others in the tops of Sitka spruces? Bald eagles. This is the last week of August, almost close season for the Boat Company. Rain and dull skies were expected, but instead sunlight turns the waters of the Inner Sound a radical blue.

Most of these guests know the philanthropist of old. Is he all right? Sure, he's all right. That damn signal stayed good for a solid half hour. He checked his stocks, downloaded his emails, and scrolled through their attachments. They've come from lawyers. These documents are part of his latest litigation. He's always in litigation. Hell, why own the damn boat if you're not going to litigate? This is fun.

'Has anybody seen Michael?' a guest asks.

'You know Michael. He's in his office.' Winsome McIntosh smiles, because she's the perfect host, but she's calculating. The 'office' is a cabin her husband appropriated the moment he found it was free. Winsome's ambition to give Michael a break, to rip him free from his data and newsfeeds and bring him out into company, was a loser from the get go. Still, she keeps hoping. 'He'll be out for cocktails.'

A skiff is lowered and a crewmember dispatched to a raft of ice that has floated loose from an iceberg. He hacks off a chunk and brings it back. Chopped into pieces, the ice is as pure and ancient as any that's ever settled on the bottom of a whisky tumbler. That might tempt Michael from his lair. He's American, but the McIntosh ancestry is what counts. 'I'm Scotch,' he'll declare, and grin. 'Not the nicest people in the world. Pugnacious.'

Imagine a boat for Michael McIntosh, and it would be this one. Called *Mist Cove* after Michael's favourite Alaskan bay, its luxury is modest. The sofas and armchairs of its lounge are like a gentleman's club, while the removable plastic windows help turn the aft into a panoramic dining room where guests settle around one large table. Forty metres long, the boat is powered by a reconditioned World War II engine and modelled on a Korean War–era minesweeper. Those wartime reference points are not incidental. Guests might be on a leisure cruise, but Michael is on a campaign.

Drinks are poured, and guests wander out to the railings to stare out at a passing cruise vessel that is even more leisurely than their own. Its passengers are seals, and the vessel is a raft of ice broken off from a glacier with a shimmer of pale blue captured inside of it. Seals and people gaze at each other.

We hear a low laugh. Michael's emerged. He's dressed for the evening in a crisply laundered checked shirt. Winsome's responsible for his shirt; he's responsible for shaving. His cheeks are hollow and graded with stubble. Maybe he laughs because he found the seals funny, or maybe the people. He has been coming to Alaska for 60 years. He doesn't need to see any more such scenes. He just needs to know they're still happening. It's good that seals hitch a ride on miniature icebergs. Good that influential people come up from the lower 48 states to catch something of the wonders of the place. And good that he's got a pack of lawyers fighting to keep it this way.

'I'm essentially a city girl,' Winsome says of herself. Brought up in a military family, she moved every two years and had to keep making friends. 'That taught me people skills. Michael and I are very much opposites. He's a very solitary guy, very happy in his own skin and company. He once told me he could come and live in a cabin out in the wilds of Alaska for the rest of his life. I'm much more people oriented. We complement each other. We have opposite talents.'

You could invent a new word for Michael. Ditch 'philanthropist', a lover of mankind, and invent 'philarborist'. 'To this day,' he admits, 'I'm more affected by the way we treat trees than the way we treat people.'

The first tree to really bond with him was a horse chestnut. It grew on the family estate, some hundreds of shoreline hectares near the town of Port Washington on Long Island. He played in its shade. There's scarce a horse chestnut left in the USA anymore. Michael was 11 when wartime gas rationing moved the family closer to their New York offices. You can't blame a kid as young as that for not saving a whole tree species, but I bet sometimes he wonders.

He was a sparky kid. He remembers chasing his piano teacher around the living room with a knife. At 80, that pugnacious part of him raises its fists in defence of an oak he had the city plant on the street outside his home in Washington, DC, to replace a small one that died. It was 2.5 metres tall when it went in, and now it's 15 metres and its branches spread. Woe betide any trucker who parks so as to bend or break any one of those branches; Michael's out on the street cussing them and calling the police. 'You can screw with some things,' he says, 'but you can't screw with my goddamned trees.'

He's fond of a spruce that grows at Sitka airport. Most wouldn't notice it. It's only one metre in diameter. What people don't respect as they fly in and out is the tight band of growth rings that reaches into its heart. They show that this tree was already growing in this spot before

Columbus set foot in America in 1492.

That tree's like Michael somehow: rooted, stubborn, slow to show itself. And then it hits you: this tree owns the territory, and you're trespassing.

It's hard to get Michael McIntosh to admit to a sentiment beside love for his wife and anger at everything. Go looking for some pattern of purpose, and you're on your own. 'Saving a million acres wasn't in my childhood,' he says. 'That came at 37, 38, 39 years old. That's when some of the principles of the work you are going to be doing for the rest of your life are established.'

Some of those principles were jarred into place in the 1950s. Michael had ideas for a career. He taught regular dancing at the Fred Astaire School of Dance in order to pay for tap dancing lessons, and danced in the chorus for the out of town tryouts of a musical. Permanent hire on Broadway was on the cards. His father got wind of it. Michael was dispatched to an Alaskan cannery. Instead of the boards of Broadway, he was treading the bark of logs strung together in the Inner Sound.

The cannery was part of the family business — one that just happened to be America's biggest retailer. Founded by Michael's mother's grandfather in 1859, the Great Atlantic and Pacific Tea Company became a household name as A&P. In 1930, it already had 16,000 stores and sales of $1 billion. In 1936, when Michael was three years old, it brought in the concept of the self-serve supermarket. 4,000 of these larger supermarkets were opened by 1950. A&P made their move into Alaska after World War I, when prices for canned salmon plummeted and they snapped up failing canneries.[1] Michael's 1951 job was at Waterfall Cannery with the Nakat Canning Company, America's largest fishery and an A&P subsidiary.

Waterfall Cannery opened in 1912, carved out of 23 hectares of wilderness on the west coast of Prince of Wales Island. It became its own community for the months of the salmon run each season, with workers arriving by seaplane or boat. Michael was set to work on the fish traps —

nothing more than a series of logs lashed together in squares with a cable running through 45 metres from one of the logs to the shore where it was anchored. He learned to handle the five metre tidal swings and drop a wire mesh net down to the bottom. Salmon tended to migrate along the shoreline, where they would hit the net and swim into the traps. The catch was between 10,000 and 30,000 fish per day.

The traps allowed few fish to escape. A diminishing resource was obvious by the late 1940s. By Michael's first visits in 1951, it was clear that the fish traps ought to be declared illegal. The federal government was enforcing antitrust laws against A&P, so plans for A&P to align with another large fishing company, NEFCO (New England Fishing Company), to lobby for a change in the law were abandoned. Such joint action might have been seen as anti-competitive.

Michael's was a union job with 16 hour days; the workers took off at 8am, and were back at the dock with fish unloaded by 11.30pm. On their one free day a week, Michael and some colleagues put an outboard on a skiff and motored off to explore nearby islands. He was a teenager in the largest wilderness area in the USA, among spruces that measured three to four metres in diameter and had been there for up to 1,500 years. There were seven million hectares of such trees. As Michael grew into his 20s, he came to see the danger of outliving far too many of them.

Starting in the early 1950s, the government worked to develop a timber industry in Alaska by giving companies 50 year contracts to cut 1,000 year old trees. The biggest trees offered the best return. Initially, companies were allowed to cut them right up to the banks of the rivers, causing silt to wash into the waters and block the passage of fish. Controls then limited logging to within 50–60 metres of a stream. Alaska was so far away from the markets that the cost of shipping the timber made it uncompetitive with areas in the south and along the Pacific coast. Great trees were turned into pulp for the paper industry. This new industry had to be subsidised.

The logging of Alaska increased through the 1960s, while Michael moved out to the Middle East to work for A&P. When his mother grew sick, he got a transfer to the bakery division in Chicago. On a return flight, a Pan Am airline stewardess called Winsome caught his heart. He started wooing. Bouquets would greet her as she stepped from the plane at airports all around the world. Michael rounds off the story adroitly. 'Met this broad on the way back from the Middle East and convinced her to join me in Chicago and lived happily ever after.'

His parents had created the McIntosh Foundation in 1949. Much of his mother's wealth was swept into it on her death. Michael entered fight mode to wrest control of the Foundation from its previous trustees. 'We had a debate,' he understates.

His parents were Republicans, which in their era equated well with being conservationists. They had given Michael a lifetime membership of the Sierra Club, founded in 1892 with a focus on protecting the natural environment of West Coast America but spreading into being a dominant national organisation. It felt comfortable to shift Foundational support to environmental causes.

The Clean Water Act of 1972 rounded off a package of environmental legislation that was like a lovechild of the 1960s. From a radical, feel-good decade, relative youngsters set boundaries for protecting the natural world in the more straitened decades to come. The USA had set the global standard. Lawyers now had the toolkit to address environmental damage. But lawyers are expensive. Michael McIntosh just happened to have a pot of money he could dip into.

'Foundation money is public money, not our money,' is how Michael states it. The Foundation had his name on it. He could imbue it with his character and make it pugnacious, but it didn't belong to him. He applied the business ethic of achieving maximum return on investment. '$10,000 here and $10,000 there on a lawsuit got you a heck of a lot more than buying $10,000 of land, so why not take a chance? Could we do more public good one way versus the other? If it worked, that was a good use of public money. Better than our Congress usually manages.'

Not that Michael restricted the Foundation to such $10,000 grants. His family foundation is a small one, a flyweight among heavyweights.

They had to pack a punch. It was in his family genes after all. Michael used to visit his grandfather Arthur Hoffman in the 1940s. The man was old and his health was fading. Even so, he brought fighters into the boxing ring in the basement of his New Jersey home and worked to give them a pasting. The lightweight champion of the world? Bring him on. The old guy and the champ slugged it out.

Michael states his philosophy. 'A small foundation, giving away $1 million a year, can give away a hundred grants of 10,000, etc., and be invited to all sorts of parties but not be doing anything. Four to five grants of $200,000 apiece, which in those days could make a difference, that's where we went. Seemed like a smarter use of money.'

The Ford Foundation had just funded the nascent NRDC's battle with the US taxing authority, the IRS. In 1970, NRDC was freed to be a not for profit organisation and litigate. Michael McIntosh matched the Ford's funding of $400,000 ($2.25 million as a current equivalent), repeated at two year intervals.

'We were different from many,' Michael recalls. 'If we liked the organisation, we gave a general support grant. That's been our history. But many foundations only wanted to fund a particular project. It didn't always cover all the organisation's costs.'

Those were heady days. Six months after joining the board of NRDC, Michael McIntosh was on the board of the Sierra Club, and would join the boards of the National Audubon Society and the Wilderness Society too. He gave out similarly large grants, spending down the Foundation's capital.

Why the hurry? In 1971, the US government was set to give a contract through its Forestry Service to wood product company US Champion Plywood to cuts tens of thousands of hectares of old growth on Admiralty Island. These were some of the trees Michael had visited 20 years before. One hundred and sixty kilometres long with 1,100 kilometres of shoreline, Admiralty Island is home to the world's largest populations of grizzly bears (some 1,600 of them, outnumbering humans by about three to one) and bald eagles. Trees would be hauled off to a pulp mill at Echo Cove just outside of Juneau. They would emerge as 8.4 billion board feet of timber.

Michael funded the Sierra Club Legal Defense Fund to bring a series of lawsuits against the US government to block them giving acreage. These forests of Sitka spruce and western hemlock first took root 7,000 years ago, as ice sheets pulled back from the region. Clear cut these forests and that's it, they will never come back. And nor will the ecosystem of which they are a part.

One of the wonders of Alaska is the interdependency of life there. When the guests on Michael's boats walk through the forests, they're trained to shout out 'Hey, Bear!', for this is the grizzlies' territory and you don't want to surprise them. These trees are the bears' home. When guests tread the forest floor, the thick mounds of moss make it like walking across a sponge. Water seeps steadily from these sponges to fuel the streams. The forest canopy provides the shade that keeps the waters cool. The stream bulges around fallen logs that offer spaces of protection for the salmon fry. Insects drift from the fertile forest to serve as food for the fish. And the salmon, of course, return from crossing the oceans to push up against the flow of these streams so they can spawn. Bears grow comradely and line the riverbanks to welcome the salmon home. They can reach into the water and snatch at fat salmon with their paws. A fair catch might be 40 such fish, which they ferry off into different parts of the forest.

No bear needs hundreds of kilograms of salmon in a day. They choose the tastiest morsels: a smidgeon of brain maybe, perhaps a liver. The rest is left to break into constituent nutrients. Fat joins with protein and nitrogen and phosphorus to ooze into the forest floor, where the roots of trees suck them up. Those fish that do manage to spawn upriver will also die. Bald eagles, mice, deer, jays, ravens, crows, insects — so many different creatures feast on the bounty to carry the nutrients further. Salmon forests deliver more wood per hectare than any other forests on Earth, trees growing tall and broad. It's remarkable. Salmon become the trees that protect the salmon, through the mediation of bears.[2]

Michael's lawsuits were buying time for Admiralty Island; the aim was to file suit after suit, and register appeals if anything lost, whatever it cost. It took more than four years: through the end of the Nixon administration and the election of Jimmy Carter to the presidency, when he declared Admiralty Island a national monument.

For a philanthropist like Michael, this victory had the same thrill as raking in Microsoft shares in its early Seattle years might for an investor. His hundreds of thousands of dollars might have bought a few acres and conserved it that way. Instead, he had funded litigation — and saved a million acres of some of the most precious wilderness on Earth. This philanthropist was hooked. Environmental litigation was where it was at.

The lifecycle of the law

James Thornton

Environmental laws have a lifecycle, like any organism. In this body of law, there are five phases.

It starts with the science.

I knew from climate scientists that if I could only do one thing for the climate, it should be to stop coal fired power stations. So stopping coal in Europe was a central focus of our work right from the start. In the area of toxics, testing regimes are important. In climate damages, the physics of climate attribution is vital; wildlife biology is fundamental to nature protection; and so on, throughout all of our work. The only exceptions are the purely procedural rules, but these too must be made to serve what the science directs us to.

The second phase is creating policy. A good example is our work on fisheries. When ClientEarth opened its doors, the European Union was revising its laws governing how fish are taken in EU waters. The biologists told us that unless the law changed radically, fish populations would crash. So our team created a new policy approach that would both protect fish and keep fishers in business in the long term. When I was starting, many Europeans told me that policy was a special realm too complex for mere lawyers to understand. From my perspective, it is just one of the domains that we work in to reach our goals.

The third phase is law making. To keep to the fisheries example, the team then took their vision and held it up against the proposed EU law. This let them see deficiencies in the proposal and gave ideas for improvement. There followed years of detailed work in the European Parliament in Brussels, supporting and helping draft a strong law.

The fourth phase is implementation. Here a law starts its life in the world. A government agency is responsible for making the law work, and for applying it to regulate some arena of life and business. The European chemicals law, for example, is designed to get carcinogenic and other dangerous chemicals out of commerce. Whether a law works or not depends on how it is implemented, and also on the final stage.

That final phase is enforcement. There will always be actors who violate a law. They may be private citizens, businesses, or governments. Every law needs to be enforced. If you pass a law and do not enforce it, you in effect authorise the behaviour you sought to prohibit. My experience in the US of bringing scores of successful enforcement cases led me to launch our EU enforcement effort.

Each of these stages is critical to creating a law and making it work. Until ClientEarth, no European NGO worked at all stages in the lifecycle of the environmental laws. Because we do, the work has great power. This is a key reason that the work of ClientEarth adds such value to the European environmental movement as a whole and why we can add value in China and Africa.

4

The Lie of the Land

It was January 2007. In midwinter California, the Pacific Ocean picked up its blue from a largely clear sky. Waves thumped onto the sand at Santa Monica. Two blocks back from the shore, lawyers went to work in shirtsleeves. They were among more than 30 staff housed in the Robert Redford Building, the new $5 million office for the Southern California branch of NRDC that James had founded.

James, meanwhile, was at a Dickensian remove.

January in Cambridge can strip the smile from even the most avid Anglophile. Siberian winds whip in from the Fens, threaten to strip the skin from the bones on your face, and turn the insides of windows to ice. The English natives were a hardy lot. While the Robert Redford Building in Santa Monica was air cooled by sea breezes which flowed through eco designed vents, this Cambridge lawyers' office occupied a terraced house with an attic conversion. The temperature kept to a steady 14°C by the simple expedient of opening the windows.

As well as an American passport, James had an Irish passport that cut through complications in settling in Europe. Home was now in a 1970s estate of semi-detached brick houses in Sandy, Bedfordshire. The Royal Society for the Protection of Birds was stationed on the edge of town, on land that was being refashioned to attract the Dartford warbler. A train station meant London was an easy commute just 80 kilometres south, if he ever found a reason for heading that way. A local bird reserve and a possible commute: such were the building blocks of his new life.

Back in his former home town of Santa Fe, New Mexico, stencilled signs were pinned above the machines in the laundromats: NO SOLICITORS. 'Solicitors' was a highfalutin term for beggars: impoverished folk who solicit money from strangers. At Gatwick Airport, James had passed through a language barrier and found 'solicitor' was now one of two options open to him. As an attorney in the US, he had prepared his cases and then carried them through to the highest courts, where he represented them himself. In the UK, a rigid hierarchy remained in place. A solicitor prepared a case and handed it over to a barrister, who has the gown, the wig, and the authority to represent the case to the judge.

As a member of the bars of California, New York, and the Supreme Court of the United States, James now had to make a choice. He could seek to join a chambers, don the horsehair wig, and emerge as a barrister, or apply to the Law Society of England and Wales for licence to work as a solicitor.

The decision was probably always clear. The US legal system may be rooted in its UK precedents, but it shed much of the older system's hierarchy when it adopted the Bill of Rights. A bewigged courtroom appearance is as arcane as a theatrical pantomime dame, and becoming a Queen's Counsel has no fit with a republican spirit. Ultimately, however, the decision on which route to take depended on another archaic division: barristers were at that point forbidden from taking anything that might be deemed to be direct action. As an activist lawyer, James's focus was on the use of the law as a tool to effect social change. He bought the requisite tomes and studied for the solicitors' exams.

He passed with flying colours. Step One.

Step Two? He set his case out for the Law Society and detailed his years of legal experience. The previous decade, he had been working as a Chief Executive Officer rather than in a law firm. The Law Society gave its response. His lack of recent direct legal practice invalidated James's entire legal career. He would need to indenture in a solicitor's office for two years.

Nothing sharpens a lawyer's talents more keenly than taking on the presiding authority in a new jurisdiction. James trained his anger at the Law Society's decision into a 20 page paper, outlining the various factors in his experience that appeared to have eluded their attention. A two year practice requirement was outrageous, he declared. Even three months would be needless, but that he was prepared to do.

They responded. They had surely taken good note of all of his points in his initial application. However, since he had brought them to the Society's attention once again in such a helpful way, they had seen into the merits of the arguments more deeply. Three months' indenture in a solicitor's office, rather than two years, would indeed suit their purposes. It was his first case in the UK, against the Law Society, and he had won it.

And so to those winds that cut across the Fens. To fulfil the three months' requirement, he settled into a boutique environmental and public law office in Cambridge. This office had a reputation for bringing those cases which had done most to further the expression of environmental law in the UK. Common law, which had been accreting for 700 years, was thus developed a further fraction so as to counter ecological crises. For James, the process was agonisingly slow. The planet would crumble while UK law accreted.

James ferried legal papers to the Courts of Justice in London. He climbed up winding wooden staircases to face a choice at the top: for the clerks, he could go left; for the scriveners, he could go right. It was amusing in one way, and desperately antique in another.

The firm worked as an occasional gun for hire for some of the big environmental campaigning groups, who did not retain lawyers of their own. It was a commercial company, so it took commercial cases, often for folk with what might be termed 'nuisance' disputes, a sense that their environmental rights were being infringed. The lawyers would sometimes find an indigent, someone who might be affected by an act of environmental degradation, and then seek legal aid in order to pay themselves to defend that intransigent's rights. 'What we need to do,' James would suggest in conversation with his new colleagues, 'in order to change the norms of how people act, so as to improve environmental protection, is to figure out strategic cases that will have high leverage and make a lot of people change their behaviour.'

He knew this method worked. He had countless instances from his American experience to support his argument. Well, maybe that's how things work in America, he was told, but in Britain lawyers should be humble. They respond to the issues brought to them by clients. James was used to the American system, where people had rights, constituted in the Bill of Rights. In England, citizens had freedoms rather than rights, patterns of behaviour set in place by ages of usage. His new colleagues felt that a written constitution was simplistic. Maybe some of their deep understanding of the virtues of an unwritten constitution would reveal itself to him by degrees?

Maybe the democratic essence of the monarchy would show itself to him too, and he would start to pore through the cricketing statistics in *Wisden's Almanack* because he had conceived a passion for the game, but it was unlikely. What he found in Britain was a very mixed system, partly unwritten and partly written: lots of statutory law in all directions, some common law, and no written constitution. For him, that lack of a written constitution meant it was much easier for the government to chip away at freedoms.

Winsome McIntosh spoke to one of James's old colleagues at NRDC, and learned that he was in England. 'No? No kidding! Give me his contact information, and I'll try and find him. It'll be fun for Michael.'

It would be a little male company for Michael McIntosh too. Winsome had brought a group of philanthropic women to London, shaking their hair loose of normal concerns to have their annual meeting in a new locale. They were part of a community Winsome had founded called

Rachel's Network, named after the environmental writer Rachel Carson. Each woman had committed to donating a minimum of $25,000 a year, up to a million plus, largely to environmental causes, with a strong focus on promoting women in leadership.

Michael had come along to London for the ride. He could sit in the bedroom and work financials on his screen, rocket off the odd legal complaint or two to protect some new patch of Alaskan ecology, and then emerge for dinner.

I headed out with Winsome, Michael, and James to the Mayfair restaurant Greens, which was something like a club, with its dark and polished wood and scarlet cushioned seating.

The dinner order was surrendered to the waiter, linen napkins were unflapped, and the wine was poured. It was time to start the conversation. Winsome was good at these events. She set the ball rolling. 'So, what's public interest law for the environment like here?' she asked.

Tables in Greens were so spaced that diners could hold a private conversation. For now, there was silence. Michael and Winsome looked at James, and waited.

'I don't know,' he admitted at last. 'I haven't seen or heard of any.'

At least these listeners from the US shared the same basic terms of reference as James.

'It's been confusing to even talk about public interest environmental law over here,' he explained, 'because there is a term in British law called public law, and public law is anything that isn't private law. So private law is contracts, and torts, and that sort of thing, and public law is law involving the government: anything in the public domain rather than the private domain. Barristers tell me, "Oh, well, I'm a public law lawyer too," with absolutely no understanding of what I am talking about.'

A dozen Colchester oysters slipped down James's throat. With the lack of public interest law groups in Europe, it looked like he would have to take a corporate law job. The planet might be hung out to dry, but at least on corporate fees he would be able to afford his own meals at Greens.

The conversation turned to the string of victories won by James and

Michael when they had enforced the Clean Water Act, now decades ago.

Those had been the days. James recalled the group of 20 medium and small sized companies in Connecticut, electroplaters and their ilk, who he had identified as some of the most toxic polluters in the North East. They received the industry newsletter which focused on James's litigation programme, and grouped together to hire the proficient law firm of Beveridge and Diamond, with its focus on environmental and natural resource law, to defend them. They did not expect to win — nobody was defeating James's law suits — but brought in the partner Harold Himmelman to negotiate a good settlement.

'Harold was a good guy, a bright fellow. Do you remember what you made me do to him, Michael?'

Michael McIntosh smiled. It was worth hearing the story again.

The group of companies made an offer through their lawyer. James did not think it was enough, but at least he could settle 20 cases at once and move on to another 20.

'I told you the amount,' James reminded Michael, 'and we just argued and argued. "Well, you could get twice as much penalty out of them," you said.'

'Was I right or was I right?'

The penalty would be paid to the Connecticut River Watershed Association. A stiffer penalty would teach the companies an even stronger lesson — and anyway, it was fun to extract the full pound of flesh. So yes, maybe James could get twice as much from those companies, something like $450,000. However, he had already accepted their offer to settle by noon on a certain day.

What if they miss that? Michael asked.

James knew he could beat them, and extract attorneys' fees for the charity. Michael had persuaded him. Yes, he would double the penalty if they missed their deadline.

'You wouldn't let it alone,' James reminded Michael. '"Will you really do that?" you said, and made me promise.'

These were the days before the internet: the offer had to be an original document and not a fax, it had to be a settlement offer signed and ready, and it had to be with James by noon. James waited. At noon, the lawyer rang to check that the offer had arrived. It hadn't. The offer arrived while James was explaining the consequences of late delivery.

'It was a brutal lesson,' James noted. A brutal lesson for the opposing lawyer, and for James too in the use of money as a tough negotiating tool.

'They paid up,' Michael added. 'When you've got the strongest hand, you don't ask to leave the table.'

Winsome listened to the stories and took note. She had been mothering her family back in Palm Beach in those days, and started a community foundation down there, but had played no direct part in the environmental actions. As Michael and James traded memories, the story moved on to Gwaltney and Big Ham, and Michael's laugh got louder.

Winsome's mind clicked away.

It was good to see her husband having such fun. Where James focuses his strategic thinking on legal manoeuvres, Winsome divides hers between philanthropic manoeuvres and attempts to shift her husband out of plans he had set himself. She has more success with the philanthropy. Why shouldn't he sell the Boat Company and spare himself the stress and expense of running a business in Alaska?

'He needed something else to keep him occupied and interested,' she later recalled. 'When our dinner came along, it seemed like an absolute natural to me to lay the groundwork and begin to do something here. Around law, something he knew, certainly a challenge, and 3,000 miles the other way. What happened is he's never going to sell the Boat Company and I began to spend more of my time developing ClientEarth. Something I hoped and planned for Michael to do, but that's what I'm doing. I don't always get it right — I do make mistakes.'

If you make mistakes, do so by degrees. That's the logic of the toe in the water, the trial subscription to a learned journal, and the small tattoo below the neckline. The McIntoshes' offer came before the pudding. 'Write us a report. We'll fund it. Investigate the true state of public interest environmental law in Europe, and let us know what you find.'

The McIntoshes knew the 1980s New York version of James Thornton. John Adams, the founder of NRDC, knew James then and also as he moved on to San Francisco and subsequently founded their LA office. 'James was a superstar,' he told me. 'We had a number of people who excelled in NRDC, and he was one of those. He was always able to go broad.'[1]

I first met James in 1991, when he stumbled into Europe in the wake of establishing that Los Angeles office. For a decade, I had joined him on occasion when meeting potential funders.

For the supplicant, these occasions are something like shark fishing in deep waters. You're normally on the funder's turf, or in a restaurant where you would never presume to dine on your own. The philanthropist is the one with the power, and a keen intelligence about the game that is being played. Conversation is a sophisticated interplay through which both sides size each other up. At some point, not too soon and not too late, comes the pitch, the big ask. What might the funder be prepared to give?

Of course, this is not like a shark hunt. You don't catch the shark. The shark, if you're lucky, concedes to flip you a mackerel or something. That thrill of entering new waters, of engaging with creatures of great power (for wealth is power), regularly has that dynamic of bringing back some catch. It's how good causes are funded.

'Boy, that was different,' I said to James after we saw the McIntoshes back to their club after dinner. It was night and so the streets of Mayfair were empty, which was handy, for I was a bit stunned. 'I've been to a lot of these meetings, but I've never met anyone more acute than Winsome. Those questions she asked: we were truly played.'

That shark hunt had been turned around. The funders had come into European waters, and they were not here to be caught. James was to be the catch. 'We've founded between ten and 15 NGOs,' Winsome would later tell me, 'and they are all still around. What's most important is investing in the right people. The programme will follow if you've got the right people in place.'

James had done his homework, aced his exams, and was able to present himself as an English solicitor comfortable working in all of Europe. Winsome recognised in him that same keen eyed lawyer that Michael had teamed up with to implement the Clean Water Act.

Of course, 20 years change a man. Briefly, here's a survey of qualities and experiences from that time, because they will intimate their way into the manifestation of public interest law in Europe.

How do you set up a law group in Europe? Well, James had set one up before, so he now had that experience to draw on. The Los Angeles office of NRDC, which he founded, still goes strong, and in term of its fundraising is the largest contributor to the group.

That also meant he also knew firsthand the intense rigours of setting up and running such an organisation. His switch to the West Coast had been propelled by wishing to train with his Zen master. Zen practice became a part of dealing with the pressures of both saving the world and forging the regional office with its own identity within an existent national group. While he does now lead meditation courses for the staff in the ClientEarth offices, the effect does not need to be that overt. Remove yourself from the need to react all the time and you create a slightly more tranquil space around you. Staff would enter his Los Angeles office and just sit for a while, to grow calm, and then they would head back to their work.

Not that James is always a calming presence. 'We both bear the scars,' John Adams told James, with a wry grin, as he reflected on the years of their working relationship at NRDC. James was somewhat shocked. He'd only been aware of his own scars.

'An hour with Mr Thornton, also a published novelist and poet, is a humbling, somewhat exhausting experience,' the *Financial Times* would write of James in 2016, and reported on James's own sense of Zen. '"Zen helps," he says. "You get a certain calming and an ability to problem solve. You study the problem, you learn all the facts, all the rules, assimilate those, and a creative solution will emerge."'[2]

James spun out of a near decade with NRDC in 1991, and took time to evaluate his life. This took the form of 14 months of spiritual retreat in Germany, during which he travelled to Dharamsala in India for an hour long private meeting with the Dalai Lama. What should he do next? How might James now apply himself to resolving environmental problems?

'You must become confident and positive,' the Dalai Lama advised him. 'And then you must help others to become confident and positive. The long term solutions to world problems, including environmental problems, can never emerge out of an angry mind.'[3]

Back in the US from that retreat in a German village, James accepted a 1992 grant from the Nathan Cummings Foundation and interviewed 50 significant players in the environmental movement. He discovered that these activists adopted anger as the basis of their work. 'It was often stated in the interviews that an activist needed his or her anger to be effective ... the activist didn't know any other basis of effective action than anger, and was worried that if anger was not present, there could be no action.' He suggested that this explained how 'the environmental movement has tended to preach in strident, negative tones'. Instead, James sought a combination of 'intimate experiences with the natural world and deep self-reflection' to help achieve 'radical confidence'.[4]

He founded the NGO Positive Futures, which brought meditation techniques to activists and policy makers. In this way, the activists still feel their anger, but learn to step aside from it. Anger becomes a source of energy, not the driving force.

Try this exercise: see yourself as so angry that you call in a lawyer. You'll vent your anger, but you don't expect the lawyer to join in. You won't rush out into the street together and shout. You expect the lawyer to ask you questions, one after another, till she's reached the cause of your upset. She will look at the pattern of your behaviour, at the pattern of behaviour in others that you find so abusive, and she will see how the law can be applied. She's on your side but she's clear headed. She sees how much you hurt. Her aim is not victory but remedy.

With the Earth as her client, the lawyer follows a similar approach.

She doesn't rush into the streets and yell as loud as she can. In effect, she interviews the client. That means taking time to appreciate what natural system is at play, and how it is being disturbed. The process examines abusive behaviour, and then it goes deeper to look for root causes. It sees how the law can be applied. This technique may not be overtly meditative, but it does involve setting oneself aside from anger and becoming at one with the Earth in at least an intellectual way.

Environmental activists, as shown in that 1992 survey, acted from anger. The movement that James helped found to bring meditative practice to activism has flourished in the USA, with funding from many foundations. Such practices are beginning to gain interest in the European environmental community.

James made one more career move before resuming a more recognisable legal career. He became the CEO of an international neuroscience research group, the Heffter Research Institute. Its aim was to configure new approaches to mental health and basic brain research through programmes of applied psychotropic research.

James wanted to mainstream this work, and funded a programme at Jung's former institute in Switzerland, because he recognised Europe as a continent where governments might authorise radical approaches with less fuss than in the USA. It is similar to his choice of a German village for the personal transformation of his spiritual retreat: perhaps some sense that a region with civil codes of behaviour stretching back centuries might be less threatened by what is new.

The Heffter Institute continues, and the research it supports has now connected with the medical mainstream in both Europe and the USA, with peer reviewed papers piling up, and popular articles in the likes of *The New Yorker*, *The Atlantic*, and *The New York Times*. Running a neuroscience institute also gave James the confidence to appreciate how the lawyers' route into understanding the Earth is through science.

A public interest environmental law group needs a body of laws with which to work. Such groups developed in the USA after the thrilling sweep of new environmental legislation brought in by the Nixon administration in the early 1970s. Europe was running about a decade behind. Nineteen seventy nine brought the Birds Directive, joined by the Habitats Directive in 1992, which established a network of nature protected areas known as Natura 2000, designed to protect habitats (both land and marine) and those species perceived as being under threat. Subsequent Directives focused on the likes of air quality, energy consumption, and greenhouse gas emissions. Existing environmental legislation in the various member states had to shift to accord with these Directives; around 80 per cent of their current environmental legislation is seen as deriving from the EU.

The European Commission focused on overseeing the transposition of its Directives into national laws. Did the Commission then enforce those laws? That was the question James needed to resolve: who was regulating the regulators? 'Nothing undermines the credibility of a State more than laws which are not applied,' lamented Ludwig Krämer, former German judge and a leading scholar on EU environmental law. He noted the Commission's good intentions, 'which are later on forgotten or bypassed ... when rules are to be applied and enforced. In the conflict of environmental interests and economic interests, in 999 out of 1000 cases, the environment loses.'[5]

Handily for James in 2005, the year he set out to report on environmental law in Europe, an analysis for the European Environmental Bureau in their *EU Environmental Policy Handbook* offered the following damning diagnosis:

Many important environmental laws from the last 10–15 years are starting to 'bite': deadlines are approaching, many objectives have still not been achieved and measures have been put in place too late or not at all. Laws are often not properly understood, badly transposed into national law and the administration ill-equipped to deliver. National authorities and courts are often unable to interpret EU law and fail to provide citizens and their organisations — and thus the environment for which they are a voice — with their proper rights. Nothing could be more devastating for Europe's acceptability among, and accountability towards, its citizens than failing to apply and enforce EU environmental law. Any environmental law which is not respected by a majority of Member States over some time risks becoming obsolete and is a wasted opportunity.[6]

So Europe had its laws, but implementation and enforcement was shoddy. How robust was the system of public interest environmental law? Was it able to rise to the urgency of the crisis? James set up a series of interviews across Europe so he could find out.

He started in Britain, but the quest proved fruitless. 'There was no non-profit group of lawyers that was working on behalf of the public, and on behalf of the biosphere, in any way like we do now.'

These were pre-Brexit days, when the ecological folly of the UK's divorcing itself from a common European environmental cause was yet to reveal itself. Even so, James, Irish by dual nationality, never conceived of anything but a pan-European operation. Following his UK interviews, in which he met with the country's environmental lawyers, he charged through the Channel Tunnel on the high speed train to Brussels, and would later carry on to Warsaw.

He met with the heads of the Brussels offices of the big European environmental organisations. They were working on trying to lobby and improve legislation. They didn't work on implementing laws, or enforcing them, and none had a practising environmental lawyer on their team.

'I knew that there was an enormous gap,' James reported. 'They didn't know that there was. All of these bright people thought, "Oh, there are lawyers involved somewhere along the way, because lawyers must be necessary, but they must be somewhere else. We don't know where they are." Astounding.'

In the United States, Professor John Bonine counted between 500 and 600 full time public interest environmental lawyers.[7] He brought his enquiry to Europe. He told James how in all the continent, including Russia, he counted just 24 public interest environmental lawyers.

'He worked hard at it,' James said. 'He's a knowledgeable guy. I actually knew many of those people over the course of a while, and I would say 24 was quite an overestimate, because a number of these people weren't actually working full time practising environmental law. Some of them were in management roles, or were in-house counsel doing contracts for the organisation and not practising environmental law. Another one, for a huge organisation, was doing land planning advice.'

The closest Europe offered to the US model of public interest environmental law was in the then Czech Republic. *Ekologický právní servis* (Environmental Law Service) is 'a civil association that was founded informally in 1995 as a student volunteer group'.[8] 'Until maybe 2000, it was really an eco consultation centre for public environmental issues,' Jan Šrytr, the group's head of the Responsible Energy Division, told me. 'One of the first really large cases was against the Nemak Corporation, which was about to build a factory in the northern part of Bohemia. There were some affected local farmers, and we helped them over several years to defend their rights.' Their work was funded from the USA and, as Jan Šrytr sees it, also 'inspired' by the US public interest law model.[9]

His colleague Kristina Sabova says that they use law as 'a tool' to achieve change. 'Greenpeace has the legal units but they do not frame themselves as a public interest organisation. Otherwise there is no-one who uses the legal force in their environmental cases.'[10]

Ludwig Krämer estimated that there were around 100 full time environmental activists based in Brussels at this time. None of these was a practising lawyer, and they were set to oppose an army of 20,000 commercial lobbyists.

Surveying the field in 2007, the French scholar Yves Dezalay noted:

'Among the NGO milieu, the connotations of the law were mainly negative.' Dezalay saw how top law graduates in the USA vied for internships with public interest environmental law groups, which later allowed progress into a business career, whereas, 'in European countries, there was an almost insurmountable barrier between the world of NGOs and that of business professionals ... This divide strongly restricted the possibilities of investment in NGOs in the legal field.'[11]

In the USA, Frances Beinecke, an early NRDC employee who became the group's president in the 21st century and is now a trustee of ClientEarth, commented on what it meant for NRDC to bring legal capacity to the environmental arena. She saw NRDC as 'bringing professionalism to the environmental movement'.[12]

This then was challenge number one: ramp up the public interest environmental law movement in Europe from a standing start, so as to level the playing field against 20,000 commercial and special interest lobbyists.[13] It would need all the Dalai Lama's putative radical confidence to take on the task.

Sadly, the dearth of crusading environmental lawyers was only part of the problem. The whole legal system, it turns out, was set up to exclude lawyers from acting on behalf of citizens.

'From my perspective, coming in,' James noted of the European legal systems, 'I found what seemed to me to be a very antiquated system, which in many ways wasn't able to deal with the needs of the contemporary world.'

One prime problem in the UK was the basic cost rule that if you bring a case and lose, you pay all the costs of the other side. 'It is supposed to avoid vexatious litigation, unless you're very rich,' James explained. 'For hundreds of years, that might have been all right. Why I say the system isn't capable of dealing with the modern world is that over the last 40 or 50 years, a certain type of litigation has emerged, which is called public interest litigation or litigation in the public interest. It is basically public spirited litigation on behalf of everybody, and that's a new concept. When the plaintiff brings a case to clean up the air or the water, they do it for the benefit of all citizens.

'You don't get to bring a case unless you get permission from a judge. If the judge says the case has a good probability of success, and you go through the permission stage, you are already acting in the public interest. You shouldn't be subjected to the risk of huge costs should you lose. The system was blind to the difference between the traditional sort of case, contracts and tort, and the modern sort of case.'

In terms of the public's rights of access to the courts on environmental issues, James saw particular problems to be addressed in Germany, the UK, and at the level of the EU.

In 2005, scholars working for the United Nations Economic Commission for Europe noted an 'existing enforcement deficit with regard to environmental law' that 'could be tackled more successfully if more extensive litigation rights existed'.[14]

James added all these facts to his report to the McIntosh Foundation.

The McIntoshes read James's report of all he had discovered, including the payment of costs by unsuccessful plaintiffs.

'That outraged us more than anything else,' Winsome recalled. 'It was basically shutting out most of the people from the judicial system, and that just wasn't right. That doesn't only affect environment, it affects human rights and education and every other NGO issue based sector that we have. It was time for some commitment from the donors.'

You invest in the person — that was the McIntosh's credo. They had had serial success in starting new charities from the same principle. They believed they had a good leader in James. Funding this time had to be modest, a secretarial level wage to see James through. In the early days, Winsome and Michael had run their Foundation so that it did not grow; it was essentially the same size as when they started 40 years before. 'We shoved as much money out to move the movement along as we could,'

Winsome said, recalling those early days. 'We can't now be the singular donor. We don't have the proportionate resources we did in the 80s, to make the same kind of an impact and do things on a bigger scale.'

That they stayed small and maintained a family board allowed them to have extra fighting spirit. Family boards can take risks: they can fail without repercussions, and experiment with social change in a way that others can't. Even on a shoestring budget.

What were we letting ourselves in for?

Relocation to London for one thing, as James knew his new organisation had to be close to power. We would have to separate for much of the year; I would have to take an academic job elsewhere in the country, as we needed a secure income. James had to surrender a corporate salary, forever, so there went any easy retirement. And this new start-up would be all consuming for years.

The move was also inevitable. James had already started four social organisations. Committed social entrepreneurs don't give up easily. He was ripe for another. His first two organisations were engaged with the application of environmental law. The second two focused on consciousness change, so that people could work on altering those behaviours that triggered ecological crises. When he scanned for new opportunities for environmental law in the US, they were constrained. He could focus on one area of environmental concern perhaps, or one region.

In Europe, he found a vacant niche. No one ran the type of pan-European law group he envisaged. In the Czech Republic some years later, the Environmental Law Service rebranded itself as Frank Bold, to include a commercial law firm that helps fund the not for profit side of the group. It moved out of exclusively environmental issues to include human rights and anti-corruption work. 'The idea of having a strong organisation in this field that will do campaign work and use law as a main tool probably is revolutionary,' Frank Bold's Jan Šrytr told me. 'It's us and ClientEarth.'

Traditional and social entrepreneurs share common traits, though studies show social entrepreneurs exhibit higher creativity, risk taking, and need for autonomy.[15]

A separate quality of entrepreneurs is that of pattern recognition. A tale of James's undergraduate years studying Philosophy at Yale amused me in this regard. He won permission to take a PhD course in the evolutionary biology of arthropods. Everyone else was a Biology PhD student. The final exam was a challenge to how well the students understood the design of everything from beetles to crustaceans. Specimens, some in bottles of alcohol and some under microscopes, were chosen to be difficult to identify. What group did each belong to? Pattern recognition! James jumped to the task, and the undergraduate outscored all the Yale PhD candidate biologists.

A study of entrepreneurs who had each started more than four enterprises shows they all engaged in an active search, which they restricted to their own areas of considerable knowledge. 'In other words, they reported engaging in a process very similar to that involved in pattern recognition,' notes Robert A. Baron, the Spears Chair of Entrepreneurship at Oklahoma State University, 'a process in which they employed their existing cognitive frameworks and knowledge to notice connections between diverse events and trends. Indeed, many stated explicitly that they had recognized opportunities by combining a number of external factors into a meaningful pattern.'[16]

So we were all set for James to engage in an active search, use his experience to deploy his pattern recognition skills, and haul a pan-European organisation into existence out of nowhere, with minimal financial reward. The commitment was a done deal. As Kierkegaard noted: 'If I were to wish for anything, I should not wish for wealth and power, but for the passionate sense of the potential, for the eye which, ever young ... sees the possible ... what wine is so sparkling, what so fragrant, what so intoxicating as possibility?'[17]

The UK lawyer Peter Roderick fronted Friends of the Earth through some of its most significant legal interventions. He had pioneered a public interest law model in Nigeria, working with in-country lawyers to tackle environmental abuses by the oil industry there. A previous environmental charity he had helped start was now defunct, and so he gave James the corporate shell to speed the new group's status as a charitable company.

James now had a company, and its HQ was a computer desk beside his bed in our one bedroom rental in London. For his corporate address, he chose The Strand, just a minute's walk from Trafalgar Square. The address amounted to a rented 10 by 15 centimetre mailbox at Mail Boxes Etc. 'That's a fine address for a start-up,' people would say on studying his card. 'Yes,' he countered, 'but it is really very modest in size.'

The legal scholar Chris Hilson had written about how environmental organisations were shy of using law, because they could not afford legal fees or risk paying the opponents' fees if they lost: 'That leaves the professionally and financially poor with protest as their only realistic course of action.'[18] James's intention was to establish a legal group which could change those cost rules and then provide other organisations with the legal acumen they needed.

The first step was to become a lone ranger, selling a story in a round of speculative fundraising meetings.

The environmental crisis was urgent. Public interest environmental law groups had a track record in the United States, and he knew he could mirror that in Europe. Cost rules and lack of standing barred citizens from taking action in European courts, and so that had to be addressed as a top priority.

James had been working to learn UK and EU environmental law with a view to using it to protect the environment, and strengthen it where needed. Winsome McIntosh decided the time was right to show the potential audience of donors where this thinking had taken him.

Through Winsome, James met with Zac Goldsmith. Zac was still running the magazine *The Ecologist*, but with a long tradition of using some of his banking fortune to support environmental causes. Jon Cracknell, head of the Goldsmith Family Foundation, took the nod from Zac Goldsmith and forwarded invitations to a lunch, afternoon seminar, and evening's dinner at the Sofitel, just off London's Pall Mall.

The audience was a dozen of the leading environmental law experts in the UK, and the foundation heads who fund much of the environmental work in the country. This was the live or die performance for James's ambitions. Convince this tough and knowledgeable audience that something new could be born, or forever hold his peace.

A good crowd turned out for the event. There was interest in the brashness of American donors and in a lawyer recently qualified in the UK — a lawyer set to advise the country's experts in how to reform the environmental movement.

Michael and Winsome spoke of their history of funding the environmental movement in the States. James spoke for more than an hour of how to increase the strategic use of law, and its benefits for the whole environmental movement. He focused on barriers to justice in the UK legal system, and applied the old joke of 'Her Majesty's courts being open to all just as the grill room at the Ritz Hotel is open to all.'[19] Only the rich could afford to defend environmental rights. James proposed a strategy to change all this, and democratise access to the courts for all citizens. It was a strong challenge to the system, and questions were robust.

Would there be support for these radical proposals? During the break before cocktails, one of the leading environmental barristers in private practice approached James. The barrister said that he had no idea how James could have entered so deeply and so quickly into the UK legal system, analysed its faults, and come up with cures. I never, said the barrister, could have done the same on the other side of the Atlantic.

Dinner was served, and Danyal Sattar took a somewhat reluctant seat at the table. 'I actually quite like doing meetings, and then I have dinner with my friends. I don't necessarily want to take my work life into dinner,' he recalled. He was Director of the environmental programme of the Esmée Fairbairn Foundation, a programme that was in effect its own mini foundation with its own set of trustees. The Foundation had an endowment of £850 million. And now that he was at the McIntoshes' table, he recognised the dinner was also a metaphor. 'They were almost saying, "Here it is on a plate. We're offering it to you. We're supporting you. We're granting funding to this initiative." So they weren't coming over here, saying, "Please pay for something we've been paying for in the States." They were saying, "We've been paying for it in the States. We're paying for it here." You go, "Okay." The money is where the mouth is.'

The McIntosh plan was to use their own small funds to attract big ones. Winsome offered her version of 'sweat equity' and flew across the Atlantic to accompany James to meetings. 'That's the best kind of fundraising,' she advises, 'having someone beside you who has given money. You are then a peer. A lot of the time, they wait to see if you're still in it two or three years later. Then it's clearly money well spent and worth the risk involved.'

At the dinner, Winsome worked the room. Lacking a million dollar cheque, you conjure what you can with a smile and a button badge. 'I've got a Barack Obama 2008 campaign badge on my desk to this day,' Danyal recalls, 'which Winsome gave me back at that 2008 dinner. It's there to symbolise that anything is possible, because if we can elect Barack Obama, anything's possible.' The badge isn't alone on the desk. 'It's sitting next to Mr Incredible. My kids gave me a little Mr Incredible, because that's what we all have to be, and one of my other children gave me the dinosaur, which symbolises what we're fighting against. That toy white cat stands for conspiracy theories.'

Besides steering people towards the dinner, Jon Cracknell led a grouping of funders known as the Environmental Funders Network. This was modelled on the Environmental Grantmakers Association in the United States, and Danyal joined the fledgling UK group on a visit to the American model. He saw what issues were being dealt with over there, and then wondered who might pick up similar ones in the UK. He recalled the maxim 'It's invented in California. Then it goes to the East Coast. Then it comes to the UK. Then it goes onto the Continent.'

As a foundation, Esmée Fairbairn is attracted to the use of 'conventional tools for an unconventional purpose', and so the strategic application of law as advocated by James was appealing. Says Danyal, 'Generally, the NGO sector lacked some appreciation of what the law really meant. As James put it: when an NGO person walks in a room and the civil servants say, "The law won't let you do that," if the NGO person can say, "Here's my lawyer. He's got a different view," the debate immediately changes. A little bit of lawyering sees through some bad excuses that people might put up for why things can be done.'

Esmée Fairbairn already supported groups where lawyers volunteered environmental advice. Now they had a full business plan for an entirely different proposition in front of them. 'The idea of somebody going, "If we can just step up and scale up the depth of our use of the law, there'll be so much more than this." That was the potential we were seeing.'

The evening concluded with Danyal offering the first non-McIntosh money to ClientEarth: a grant of £20,000, the top amount that could be given without board approval. James subsequently went and presented his plans to the Esmée Fairbairn trustees. 'That's such an American approach to life,' Danyal offered afterwards, with his wry smile. 'In the UK when we see a big problem, we have a committee meeting, agree it is a big problem, and schedule the next meeting a year later. The American approach is altogether different.' He puts the approach into words: '"It's impossible! Let's do it!"'

Here's another impossibility. You move onto someone else's turf, say that you want to provide free legal help to empower their efforts, and everybody is happy.

It didn't happen.

One difficulty was the nature of environmental law within European NGOs. From his survey of 2007, Yves Dezalay noted how 'NGOs could only count on the assistance of young activist lawyers, who had very little experience and credibility in the courts' and 'found it difficult to build sophisticated legal argumentation acceptable to courts'. Because of this, Greenpeace staff in London told him, they used the courts to gain media exposure 'to turn their activists into "martyrs to the cause of the environment"'. So this was the secondary, supporting role assigned to lawyers: to boost the sacrificial profile of campaigners. 'By making these individuals into public martyrs, we could help build a platform for our cause,' the Greenpeace staff said. 'It gave us huge coverage on the media and we gained a lot of support.'[20]

I asked Danyal Sattar how well James's alternative approach had gone down with the environmental movement in the UK. 'When James first started up, he was really fired up by the climate change side of things, and rightly so,' Danyal told me. But climate change receives less than 2 per cent of philanthropic dollars,[21] and was a contentious area within the Esmée Fairbairn Foundation. The Foundation was developing its interest in marine conservation. Had James looked into that? 'I think that then took him into marine conservation organisations, and the World Wildlife Fund (WWF) had an in-house lawyer in that area. We were funding a campaign for a Marine Bill and a Marine Act. So, maybe, the odd toe was stepped on, and some elbows were around that, but not in a bad way.'

I pushed him further.

'Yes, there were a few ruffled feathers,' Danyal admitted. But it wasn't like James was positioning ClientEarth to take over the territory of others. 'It wouldn't have been like coming in and saying to Friends of the Earth, "By the way, we're going to set up a whole environmental campaigning organisation with local groups at the grass roots that's going to tread on your toes,"' Danyal said, 'or saying to Greenpeace, "We're going to build a couple of ships to go down to the South Atlantic, and compete to get in front of the whalers' harpoons." There was just such a lack of legal capacity.'

To learn more about the effect of this new kid on the block, I went out to a worker's café near London's Pentonville Prison and met with Phil Michaels.

Phil was general counsel at Friends of the Earth UK when James went on his round of scoping interviews, and soon after expanded his group's legal provision by taking on a second lawyer. 'Work that ClientEarth started doing was felt to be invading the areas of a lot of people who had been working in the field together,' he said of ClientEarth's arrival in the country. 'It was inappropriately transporting a foreign model into the system without understanding the local soil.'[22]

Yet Phil was still clearly conflicted. 'James put in a lot of work trying to understand the scene,' he admitted. 'He had as good a picture as anybody. His work pushed at the boundaries of English reticence — for good reasons, and for a lot of institutionally self-centred reasons. There was a clash between US and UK culture — in the UK, people don't say what they think. There was a lot of misunderstanding.'

And Phil Michaels saw that the UK community was ripe for some challenge. 'On the negative side, there was a cosy, slightly clubby group of people from private practice and the not for profit sectors, working together and staying together.' However: 'On the positive, they were a group of people working together for many years, sometimes institutionally for decades, building up public interest environmental law from a state of nothingness.'

He accepted how environmental law in the US was lawyer led, with the lawyers combining litigation and direct campaigning. Personally, he was more comfortable with the UK model. 'In the UK, it is campaigner led. Campaigners ask lawyers what they can contribute to the wider objective.'

As an example, he told me how Friends of the Earth led the process towards the UK's world leading Climate Change Act through parliamentary campaigning. This was legislative campaigning. As head of legal for Friends of the Earth, Phil was hardly ever called on. Ego wise, he admitted to the occasional pang, but in truth he felt the group made the right call. Unless legal input was carefully managed and controlled, he felt it had the capacity to slow things down.

For James, it was a bit of a puzzle. Through ClientEarth, he was offering free legal capacity to the whole environmental community. Why the hostility? Why didn't they all embrace it?

For comparison, I took the question to John Adams, the founder of NRDC. What was it like back in the early 1970s, when NRDC was starting out? Did the US environmental community welcome this sudden burst of legal expertise?

John tipped his head back and laughed at the very notion of his group being welcome at the beginning. 'Lawyers would ask to see our licences,' he recalled. 'It was offensive.'[23]

ClientEarth's best course of action was to secure their base and to model their own future. They were in it for the long term.

One floor up in a courtyard in Hoxton, the freshly hip part of East London: date stamp 2008. In the square of the main room, desks tuck around each other like otters. James has the small front office, whose door closes but lets sound through. Potted plants squeeze in between books. ClientEarth's London team of five have each framed their favourite photo of the natural world and hung it on the walls. There is by now also a Brussels outpost. This is pre-Skype, so, as the founding employee in the Brussels office, Anaïs Berthier is connected by a landline. The phone sits beside a ClientEarth nameplate on her rented desk inside the WWF Belgium office.

These few ClientEarth employees are the new kids on the block. The upstarts. The ClientEarth board is beginning to add lustre as word of the group spreads. Emily Young, one of Britain's foremost sculptors, chisels at ancient stone till powerful beings emerge. She sees them outlasting civilisation as we know it, so they will stare at unknown worlds millennia from now. She was one of the very first to align herself with ClientEarth's mission to protect the planet. Her friend the musician and artist Brian Eno was pulled into dinners at her Notting Hill home, to talk with James. Brian joined the board. Soon James and Brian were off to studios in Primrose Hill, where Brian was producing Coldplay's latest album. The members of Coldplay all became ClientEarth patrons and supporters.

The big players among the environmental charities — WWF, Greenpeace, Friends of the Earth, etc. — had 20th century origins, back when they were bright ideas that sparked debate around kitchen tables. This new group is manning it up in the 21st century. Decades of environmental activism has kept the fight alive. Humans know what they're doing to the planet even while they ratchet up their destructive actions. Calamity is outdistancing environmental activism. James has one tool to apply in order to redress the balance — the strategic use of law.

James laughed when I told him Alastair Campbell's analysis of what makes a winner. People who win all start from the same place: they hate losing. You've got to hate losing more than you love winning.[24]

'Winning is good, but it's always temporary,' James explained, and smiled. 'Losing is always permanent.'

Winners have an objective. James has three main ones: increase access to justice on environmental issues; limit the effects of climate change; protect biodiversity.

To achieve their objectives, winners deploy a strategy. 'They never get confused, or diverted,' is Campbell's take. 'They make their own weather, they don't just react to other people's strategies.'

Strategy is the main key. When Vladimir Putin walked into the G20 summit at Brisbane, Campbell recalls how the Russian 'told other leaders he was the only one in the room with a strategy, and that they were all tactical, adding: "You think your tactics will bring me to my knees, but you will be on your knees first."'

In terms of strategic priorities, access to justice was fundamental. Without that, the other goals were non-starters. Lawyers need laws and the right to apply them.

'The only thing that keeps our society from falling apart at the seams is law,' is how James sees the bigger picture. 'In early human tribes, there was never anything other than constant war and threat with immediate neighbours. We have so many warring factions now, that law is the structural glue that holds human society together.

'It is only gravity that holds the solar system together, or otherwise the planets would tear apart and self-destruct. Law is the gravitational system that keeps human societies moving in a concurrent direction. How is law designed? Law is basically a system of mutual restraint, mutually agreed upon, mutually enforced.'

That, of course, is an idealised view. Law is not always equitable. The whole concept of public interest lawyers taking the Earth as a client is to assert planetary rights. Resources are not there for plundering, and the atmosphere is not a free dumping ground. Arrival in England with its cost rules highlighted a huge flaw in the practice of law in that regard.

In a private case, a plaintiff seeks to advance either their monetary position or reputational position against a defendant. That plaintiff has a lot to gain personally, to the exclusion of everyone else in the world. In public interest litigation, the plaintiff is bringing a case that she, or the organisation, does not stand to gain from in any way different from all other similarly situated citizens. Fight to clean the air in a city, for example, and everyone who breathes that air benefits equally. Such public spirited litigation on behalf of everybody is a new concept.

'Until very recently the British system still penalised you with punitive potential costs, even when you were acting in the public interest,' James recalled. 'So one of the first things I did was to challenge that.'

Delegates from a whole range of countries flew to Denmark's second city, Aarhus, in June 1998. Their job was to add signatures to a document that had been argued into detail over the preceding months. This UN Environmental Convention had three principal strands: access to information, participation in decision making, and access to justice. Thirty nine signatories pledged their countries to its principles and so created a fine souvenir — the Aarhus Convention.

Such conventions can have the nature of a jamboree. Everyone wants to join, and it's churlish not to sign. Exchange contact details on holidays, and you don't truly expect folk to come knocking at your door and moving into your spare bed. Treaties signed on overseas jaunts can have the same flavour. Signatories go back home and forget about the details.

Countries took their time in ratifying the Convention. Liechtenstein and Monaco are still holding fire. The UK and the EU managed a laggardly 2005. Germany stumbled to ratification in 2007.[25] Ratifying a treaty, of course, goes no way to making you compliant.

European environmental groups had been active in the negotiations which set up the Aarhus Convention. As soon as the UK and EU ratified the Convention in 2005, a letter of complaint about the UK's 'loser pays rule' headed to the European Commission. It came from CAJE, an alliance of environmentally minded organisations in the UK.[26]

Silence ensued.

Why I do this work

James Thornton

People often ask why I do this work. The answer goes back to a lifelong love affair with nature.

As a boy growing up in New York, I would comb vacant lots for their insects and spiders, and tramp local marshes to observe birds and other enchanting beings like snapping turtles.

Entering my teens, I was the treasurer of the Junior Entomological Society, the young person's offshoot of the august American Entomological Society run out of the Museum of Natural History on Central Park West. Among the senior division's members were heroes I was able to meet, such as Alexander B. Klotz, the author of the field guide to North American butterflies.

My first mentor was a great woman entomologist there called Alice Gray, who deserves her own story one day. Under her guidance, I could have become a biologist. This would have allowed me to align my daily work with the boundless yielding secrets of the natural world. At university, I studied Biology and Literature and nascent Computer Science, while majoring in Philosophy.

By the end of my time at Yale, I determined that Western philosophy would not yield the meaning of life, and turned that quest towards Zen and its better answers, or practices.

It became clear that the alarms sounded at Earth Day in 1970 were real. If I became a biologist, I would study the living world I loved and catalogue the damage. I determined to study law. Not because I was attracted to it but because I instinctively understood it would allow me to fight on the side of life.

I was pre-adapted to studying law. My father was a law professor. He was also a master of the Socratic method. There were four boys in the family, myself the third. All became lawyers, some after detours in career. Because there were five years between each of us, the disputations around the dinner table were a good testing ground for the less experienced advocates against the more. My father acted both as provocateur and judge, keeping score for winning arguments. I fiercely loved the game. By the time I got to Yale, I also realised that I had been trained in a dinner table regime that meant I would never be uncomfortable in an argument.

In my third year of law studies at New York University School of Law, I was editor in chief of the Law Review. One of my editors, called Fred Harris, came in one day and said, 'James, you have to do the clinical programme with NRDC. They are environmental lawyers. They are brilliant and eccentric and I think you'll fit right in!' With a recommendation like that, who could resist. NRDC was only nine years old when I started my work there, about the size and age that ClientEarth now is. That taste of law practice for the Earth stayed with me. It showed me the way I wanted to use law.

The other question I am often asked is why I am in Europe. A *Financial Times* journalist yesterday asked in an interview, 'so why are you here, isn't the UK a small patch to work on after the huge scale of the United States?' The scale question is simple. I never saw ClientEarth as being confined to the UK. For me, the UK is one local jurisdiction, like California. From the beginning, I saw ClientEarth as spanning the 28 countries of the EU. And I always felt that once established, it would reach out to China and Africa.

The question 'why the UK at all' is easy to answer. My partner is Martin Goodman. He is a writer and professor, and the lead author of this book. He lived with me for seven years in the US. But human rights laws were better in Europe. It was easier for me to be legally present in the EU than him in the US. The US made him leave every six months and stay out for three months at a time.

Though Martin was able to go to our mountain retreat in the Pyrenees to write during these times of absence, the charm of enforced separation wore off, even in French. You might say I'm in Europe as a refugee from the laws of my birth country. The good news is that the United States Supreme Court has since then ruled in favour of marriage equality, something I did not expect to see in my lifetime. So on this front, the law is moving in the right direction. Nevertheless, it was the difficulty of us being together in the US that led me to Europe, and that is what let me create ClientEarth.

As a boy I loved nature. Through practice, I've grown fond of people too.

My goal is systemic change to protect people and nature. The most efficient way to achieve it is through law. This use of law to create systemic change is the most complex and rewarding enterprise I know. Every day, you ask yourself how to go deeper, how to use these powerful tools better. Once you get a taste for it, you won't want to do anything else.

5

An Air That Kills

James was quite clear: the move to London would shorten his life. He also saw no choice. You could not start a public environmental law group outside of the capital. Such a group had to be near the seat of power to influence the lawmakers.

Ostensibly, the move showed no hardship. We rented an apartment from actor friends who lived just down the street. West Hampstead was ramping up its chicness, with delicatessens nudging out the charity stores on the High Street. A short walk away were the acres of Hampstead Heath, where personal trainers led folk in pantsuits, and pedigree dogs roamed in staggered and obedient packs. The trouble was that to reach the Heath we climbed a steep hill, so engines revved and traffic belched diesel as it passed us.

Hampstead Heath was, for weekends, a touch of urban pastoral — up Parliament Hill for a blast of wind, around the duck ponds, and home again. The weekdays were the real killer. Mapmakers set two traps on the way to an underground station. One way was along the through route of Finchley Road, where buses snarled to a standstill and then roared on in a broad fug of diesel vehicles. The other way saw double decker buses caught in the tight confines of West End Lane. James returned from his short daily commute and his voice was squeezed to a croak, his lungs were burning. He brought out his new inhaler and wheezed himself back to a functioning state.

Maybe he would live for two years less. That was his best guess for how much London's dirty air would shorten his life. Science suggests he was being optimistic. Figures for 2010 showed a combination of NO_2 and $PM_{2.5}$ pollution caused the loss of 140,743 life years, equivalent to the deaths at typical ages of 9,416 Londoners.[1]

James was an environmental lawyer. ClientEarth needed to establish which issues it would fight for. Why not make clean air a priority?

He'd love to, James said, but potential costs were vast, and who would fund it?

First, he needed to change the legal systems of Europe.

Essentially, ClientEarth exists to give the environment a legal voice. A legal voice has no power if it is excluded from the courts. James's first actions were therefore brazen: In the UK, he set out to change the cost rules. In Germany and at the EU level, the matter was one of standing: rights had to be granted for citizens to bring serious environmental concerns to the courts.

Since the Aarhus Convention has public accountability on environmental matters at its heart, a structure was set in place to ensure signatories practised what they had signed. The Aarhus Convention Compliance Committee is formed of legal scholars and practitioners from among the participating countries, who sit for three year terms. They consider complaints, and when deemed necessary call concerned parties to hearings in the UN Headquarters in Geneva.

This committee is a remarkable creation. It is not a legal court, but a group vested with the power to call governments to answer for their actions. When it came to environmental issues, a country failed if it did not support a judicial system that gave citizens the right to take on a player with much more power and financial muscle than themselves. These international legal experts could call a government to account for acting against the interests of its own citizens by leaving a judicial system unbalanced in favour of corporate or governmental interests. However, a national government is a mighty force to take on.

Arm a keen eyed group of lawyers with the law, and they can beat an army. That concept is still new in Europe. Back in 1979, geographer Timothy O'Riordan compared his experience of the nascent public interest environmental law groups in the US with the current situation in Britain. 'Whereas the British seem to prefer peaceful compromise through orderly consultation,' he noted, 'the Americans prefer adversary politics where argument and counter argument is laid out before some arbitrating individual or tribunal.'[2]

While that basic distinction stays sound, James challenged the 'adversary politics' phrase when I read it to him. He preferred seeing the 'American' approach as professional, direct, and transparent, and posited the state of environmental jeopardy the world has entered as proof that the old methods have not worked.

For his bid to open courts to citizens with valid environmental concerns, James saw that the one open 'tribunal' was the Aarhus Convention Compliance Committee. Governments were not used to being held to account by a group of legal scholars and practitioners who convene overseas, prompted by a small charitable body such as ClientEarth. That committee needed to flex unfamiliar muscle if it was to open the doors to European courtrooms. ClientEarth's complaint had to present them with all the strength they needed.

It was make or break. The law group's ability to mend the planet from its European base depended on it. The lawyers decided to base their complaint against the UK on a particular instance where the government was in breach of its Aarhus Convention obligations. They went in search of such a case.

Bob Latimer lived on England's North Sea Coast, and might as well have lived in the sea itself. There are balmy days, of course, but even in summer a wind can skim the foam off waves and numb you to the bone. Stormy days wash away all distinction between the land and the sea. Bob's father built a garage out of pebbles and sand from the shore.

Bob, an engineer, converted it to a fishmonger's shop for his son, who came back from creel fishing in Scotland to run it. Bob had long been concerned about sewage spilling into the sea. As his son's shop opened in 2004, he became alert to a new threat.

Sixty thousand tonnes of sludge dredged from Tyne Docks, containing such toxics as arsenic and mercury from shipbuilding, had been dumped at sea and capped with sand and silt. The strong movement of waves in the region brought doubts that this cap would hold the toxics in place. The venture needed an environmental impact assessment, Bob believed. And the whole operation required regular monitoring.

In 2004, he sent a letter to the OSPAR Commission, the body which has responsibility for conserving the Northeast Atlantic. It was the first stroke in what would become years of correspondence to various official bodies.[3] Essentially, this was one retired engineer acting against the Port of Tyne Authority. So in one corner, there was a retiree with a madcap compulsion to keep the sea as clean as he could. In the other corner was the second largest trust port in the UK after Dover, one which employed 450 people and had just invested £11 million in new infrastructure. A company that size can take the risk of racking up legal fees, especially when such high fees work to fend off legal challenges: if Bob took a case for the public good against the Port Authority and lost, current rules would leave him bearing their costs as well as his own. To make things worse, the chances of winning such a case were severely hampered by lack of access to documentation, and by tight time constraints under which he was entitled to bring any action at all.

The Marine Conservation Society alerted ClientEarth to Bob's concerns, and the law group chose it as the primary focus for their complaint against the cost rules in UK courts. Their complaint was first filed in 2008, under the names of ClientEarth, the Marine Conservation Society, and Bob Latimer, while a range of NGOs from across Europe attached their names in support.[4] The claim held the UK government to account on four issues, the predominant one being cost. 'Not only is it prohibitively expensive to bring cases in the UK,' James said at the time, 'the financial risk of losing a court challenge and having to pay the opposition's legal expenses can amount to £100,000s. Until the UK makes the legal system accessible and fair from a financial perspective, citizens and many organisations are in effect denied their right to raise legitimate environmental concerns in court. This is in direct opposition to the intent of the Aarhus Convention.'[5]

Sheep graze the 46 hectares of Ariana Park, kept from roaming through the rest of Geneva by a security fence. This is international territory, home to the Palais des Nations of the United Nations, which after Versailles is the second largest building complex in Europe. In September 2009, groupings of lawyers made their way through marble halls to a committee room large enough to stage a ball. Dark wooden benches for the members of the Aarhus Convention Compliance Committee were raised on a platform, while other interested parties gathered around the tables below.

The UK government had dispatched James Eadie QC to the case; newly appointed as the First Treasury Counsel, and known in the profession as the 'Treasury Devil', he was the barrister to whom the government first turned for major pieces of advice and litigation.[6] He brought his own legal staff, while ClientEarth was represented by Stephen Hockman QC, working on a pro bono basis, supported by James and another ClientEarth lawyer, Sandy Luk.

A great deal of legal debate is consumed by the precise meaning of language, very often focused on a single word. In this instance, that word was 'prohibitively'. The Aarhus Convention required that legal costs not be 'prohibitively' expensive, without fixing further clarity. The ClientEarth team showed how average costs for a case were a multiple of average income. Was £40,000, say, prohibitive?

For the government, James Eadie asserted that 'costs that are merely "expensive" are permissible; providing the costs to the losing party are not prohibitively expensive'.[7] Consider having your day in court, and having to pay £40,000 in costs to the other side. It might be dear, Eadie said, but it is not prohibitively expensive.

Committee members are drawn from across all signatory countries, on a revolving basis. It took a moment for the multinational panel to understand the very British use of the word 'dear', meaning expensive, and then to translate £40,000 in costs into their local currencies. When these steps were completed, it became clear from their reactions how prohibitive UK costs appeared from the perspectives of Poland, Kyrgyzstan, and other signatory countries. The committee asked tough questions about how such punishing sums for ordinary citizens could be anything other than 'prohibitive'.

Stephen Hockman had a pile of documents 35 centimetres thick before him on his table, neatly bound in pink ribbon. He did not need to refer to them for his rejoinder. Its details belonged to a case he had represented himself, in which a householder was objecting to noise from a neighbouring factory. The hearing took two days. The householder won, but the defendants subsequently shared what costs he would have been liable for had he lost: a sum of £1 million. The householder was a businessman so enraged by the noise that he was prepared to 'bet the house' on the outcome — for selling his house would have been the only way he could have cleared the debt.

ClientEarth sought one way cost shifting, as was allowed for environmental cases brought in the USA: plaintiffs get all their costs paid if they win but pay none of the defendants' costs if they lose. That proved a step too far for the Compliance Committee, who took a year to deliver their findings.

They required the UK government to bring in 'practical and legislative measures' to make sure that costs in environmental cases are 'fair and equitable and not prohibitively expensive', and also review its rules on the time frame allowed in which to bring claims for judicial review, so as to make them fairer.[8] At least plaintiffs would no longer be liable for the costs of the opposing side.

'It's a tremendous victory for the common man,' Bob Latimer told his local newspaper. 'It will make it easier for people to bring legal action against big corporations. In the past, companies have known high legal costs prevent most people from taking cases to court, going the full distance, and many firms have played on that fact. This decision should change that situation. It has been the best part of two years since we started this, but I was always confident we were going to win. Our legal team were fantastic.'[9]

Other groups had debated the merits of ClientEarth's approach. Jan Šrytr recalled the Environmental Law Service's (now Frank Bold's) view: 'At the time we were discussing whether to go into the case or not, we had a different opinion to them, and it proved that they were right, so they won the case and it was successful and beneficial.'[10]

For James, the UK government's rhetoric could be reduced to a simple plea. 'We sign such international treaties, it's what one does to be neighbourly, they represent a jolly good ideal, and then we go home and forget about it. National governments don't exist in an ideal world. It changes the rules of the game if you then force us into a corner and demand that we comply.'

Now James had forced compliance, and it was time to make use of it.

Members of the Aarhus Convention Compliance Committee would fly to Beijing in future years, and in meetings with top Chinese policy advisers they described this ClientEarth case as their greatest act. The same committee would rule in ClientEarth's favour on their two subsequent cases, against Germany and the European Commission.

Let's slow down for a second. James's first victory in the EU was against the UK Law Society. The second was against the UK, the third against Germany, and the fourth against the EU, these three in the Aarhus Convention Compliance Committee. The first victory allowed him to practise law, the next three were designed to let citizens have access to justice, and to show that a small organisation with few resources could use the law to take on the most powerful forces in Europe, win, and make them adapt their behaviour.

Those victories changed legal systems. For the lawyers, though, it was like running a marathon in order to arrive at the starting line. In the UK example, they now had the right to bring an environmental case to the courts, provided a judge found it meritorious, they could bear their own costs, and they could risk £10,000 to the opposing side if they lost. For such a right to be substantial, they had to actually bring a case, and win it.

Ironically, air pollution in the UK increased because of steps to prevent climate change. In 2001, the government lowered vehicle tax on diesel cars because at the time they emitted less CO_2 than petrol cars. Sadly, they also run at such a high temperature that they burn the nitrogen in the air, creating nitrogen dioxide (NO_2), a poisonous gas. Besides the NO_2, diesel also pumps out particulate matter (PM), an invisible cloud composed of elemental carbon and the likes of sulfates, nitrates, and metals. These particulates can be 100 times thinner than a human hair. We breathe in the gas and it strips away the linings of our lungs. The fine particles can even penetrate the blood brain barrier to damage our brains directly.

The government knew what choice it was making. Back in 1993, it convened a Quality of Urban Air Review Group, which issued a report, 'Diesel Vehicle Emissions and Urban Air Quality'. The Department of Environment noted how 'the impact of diesel vehicles on our urban streets it to be viewed with considerable concern unless problems of particulate matter and nitrogen oxides are effectively addressed'. It was clear that 'an increased market penetration of diesel cars at the expense of three-way catalyst petrol cars will on balance have a deleterious effect on urban air quality'.[11]

A very senior civil servant in the Department of Health recalled the conflict between health issues, a significant factor in the government's discussions, and the need to be seen to be addressing climate change by lowering CO_2 emissions. 'We did not sleepwalk into this,' he said. 'To be totally reductionist, you are talking about killing people today rather than saving lives tomorrow. Occasionally, we had to say we were living in a different political world and everyone had to swallow hard.'[12]

The UK government published guidance on NO_2 levels in 2004, in which it claimed 'the UK Air Quality Strategy aims to achieve its objectives earlier than the EU has set'. Their optimism was based on old figures that showed a 37 per cent drop in NO_x emissions in the decade up to 2000. (NO_x is a combination of the colourless gas nitrogen monoxide (NO) and the brown gas nitrogen dioxide (NO_2), both of which come from burning fossil fuels.) They expected a further 25 per cent reduction by 2010. The UK Air Quality Strategy therefore consisted of a hope that things would stay as they were. A principal danger the government foresaw was the 'increased sales of diesel vehicles', which of course the government's separate 2001 policy to boost the sales of diesel vehicles duly brought about.[13] 'In hindsight,' a major report of 2016 would declare, 'the shift from petrol to diesel vehicles over the last 15 years has been disastrous in terms of its impact on air quality and health.'[14]

What were those objectives set by Europe? These came from the European Air Quality Directive, which set 'limit values', the permissible quantity in the air for sulphur dioxide, nitrogen dioxide, lead, benzene, and carbon monoxide, as well as particulate matter. The latter was divided into two categories by size, PM_{10} being tiny and $PM_{2.5}$ much tinier still. Deadlines were set for compliance, which the UK government among others steadfastly failed to heed.

This was a law with enforceable numerical standards. Accurate monitoring would show whether a government was acting in violation of the law or not. The UK made a woeful error in anticipating NO_2 reductions while lowering tax on diesel vehicles, so extreme breaches of NO_2 concentration limits seemed inevitable. The situation offered the same opportunity for enabling citizens' rights as the US Clean Water Act had done for James back in the 1980s.

Even so, ClientEarth still received minimal funding. James might risk the £10,000 cost of taking a case to court, since he was confident of victory, but he needed at the very least to hire a full time lawyer to assign to the task of taking on the UK government. Who on Earth might help back such a move?

I strode across Hyde Park. In terms of taking clean air exercise in London, this May day was about as good as it gets. An Egyptian goose huddled down on the concrete bank of the Serpentine, with her brood of goslings collected in the lee of her body, and she faced the brunt of Arctic winds. Football players gathered beneath trees while rain slashed the fields. I turned into Knightsbridge, walked beside the traffic that grows thick beside Harrods department store, and then tucked myself into Simon Birkett's mews office, from where he directs the Clean Air in London Campaign, in this exclusive enclave of the city.

Simon Birkett graduated in civil engineering, and built a career as a banker in the mergers and acquisitions section of HSBC. That gave him experience as what he terms a 'barrack room lawyer', drafting documents with prospectus standards for formal verification and the like. He ran large teams at work, and was soon gathering together residents and business groups in the city. Their first act was to take Westminster Council to court over rat running, cars that raced through an area's side streets with all their danger and noise. And then he turned to pollution.

The topic was in the air, you might say, in his childhood, when he caught on to the Tom Lehrer song about it:

Pollution, pollution!
Wear a gas mask and a veil.
Then you can breathe,
Long as you don't inhale!

Studies showed him the serious consequences that air pollution posed to public health. Beyond that, he came to see that air pollution might have some legally binding limits set to it. An early letter to the

EU Environmental Commissioner helped set him on track, for a reply directed him to focus on the Air Quality Directive.

The group Clean Air in London was constituted, and lobbied to have its concerns reflected in Europe. The Ambient Air Quality Directive, that was to be adopted by the EU in 2008, was open for input, and Simon began to feel that his arguments had weight.

Setting legislation in place, of course, is one thing. Enforcing it is quite another. Clean Air in London found it had the field to itself. Simon had marvelled at the 'fabulous' job done in the area by Friends of the Earth, but then that group switched from air pollution to focus on the Climate Change Bill. Simon spoke to the CEO of another of the UK's top environmental organisations, and asked what was being done about enforcement of the air directive. 'Well once the laws are in place, we expect them to be complied with,' the CEO told him. This has been a fundamental problem for environmental action in Europe: groups work hard to bring laws into being, and then back away before those laws have been enforced.

Simon knew he needed a partner with a different mentality. He found one, which was why I sought him out. He was returning to his office as I approached. We trod the cobbles and reached the front door at the same time. He let us in, wiped the rain from the dome of his head and his round spectacles, and led me upstairs where we settled in comfortable armchairs. He told me about the alliance between his campaigning group and ClientEarth, a merger that was primed to make legal history.

The Air Quality Directive, brought into force by the EU in 2005 and the UK in 2008, set a date for compliance of 2010. However, the UK government showed no sign of interest. 'By Spring 2009 I was wondering, what else could we do?' Simon reflected. 'Where else do I start? Because I thought the government isn't even trying to comply. They still could have applied for a time extension until 2015. I was desperate to find the right opportunity to get this problem fixed.'

London was set to hold the 2012 Olympic Games. That was the initial pressure point for Clean Air in London, and by 2010 the Olympics were already perilously close. That was when a lawyer introduced Simon to James.

'James was the first person I came across who basically got it. I was very struck by James's track record in the States enforcing the clean water laws. It struck me how wonderful that was. It was exactly the sort of vision, commitment, everything that I had been looking for.'

Simon characterises himself as an interventionist, so what he sought from ClientEarth was some sort of legal intervention. He had used Greenpeace's go-to London solicitor on an ad hoc basis to write some stiff letters, warning the Olympics board against breaching the air directive. Beyond that? 'Well, I didn't know how to begin to take the government to court. I didn't know how to get these laws enforced. I really had no idea how that could be done.'

He seeded the work with a donation, and put half a day a week into liaising with the young lawyer James set on the task, Alan Andrews. Two early attempts to force the government's hand, one in the High Court and subsequently the High Court of Appeal, failed.

'It was demoralising,' Simon remembered. 'The judge didn't seem to engage with the issues. I appreciate courts don't like getting involved in telling elected politicians how they have to do things. I feared that ClientEarth would just give up after those two cases. It is a great testament to James's character that he wasn't giving up even after those two losses. He was absolutely right because these laws are black and white. They are there to be enforced, but I think most people would have given up.'

One reason James chose not to give up was the fundamental change that happened to the UK legal system in October 2009: the role of the Judicial Committee of the House of Lords was disbanded, and its duties were taken over by the new UK Supreme Court. The highest court in the land was no longer sitting in Parliament. The likelihood of its holding the government to account was therefore greatly increased. This was an opportunity for the Supreme Court to exercise its environmental muscle for the first time.

As one of the most august assemblies in the land, the Supreme Court has done a valiant job of making itself accessible to the public. It's close to Parliament, just across from the tourist hotspot of Westminster Abbey, so family groups on a day out tend to appear for a while and then go. Salient facts about the day's cases are printed for them as handouts. All the judges are lords, but for civil cases they enter the courtroom in dark suits rather than their horsehair wigs and ermine robes. The courtroom rises to its feet at the judges' entrance, but with a sense of this being a more open era for justice.

Dinah Rose QC led the day for ClientEarth. In pollution terms, this day was marked as moderate to high. 'I assume your lordships have all been on the maps and checked the local monitoring stations,' she began. 'You are urged not to undertake strenuous activity and to use your usual medication, including asthma inhalers.'

'We decided not to play tennis this morning, if that's any help,' a judge quipped back.

'Probably a wise move, my lord.' And the show had begun. It's combat by argument, each barrister tackled in turn by the assembled lords. The whole day is a demonstration of intense referencing skills, the lords' focus targeted at evidence to be found in volumes of assembled documents.

Dinah Rose, dressed all in black with a long black skirt, also drew on more emotive facts, such as the annual deaths of 29,000 people in the UK from breathing particulates and that the effects of air pollution fall disproportionately on the poor. Walk your child to a school close to arterial roads, and the exposure to pollution means their lungs are likely to grow less fully, and therefore function less well. They are also rendered more prone to asthma, diabetes, cognitive dysfunction, and stunted neurodevelopment.[15]

The Court of Appeal had ruled against ClientEarth, saying that a mandatory order to require action from the government 'would be to trespass on the political sphere'. A judge dispatched this argument early on in the Supreme Court hearing, because 'the political judgments have already been made in this case by the European Parliament and the European Council in deciding to set a particular limit value and to make it mandatory and to require it to be implemented by a particular date. That is the political judgment.'

Dinah Rose summarised the government's position as 'it's sufficient for us to admit we're in breach and to say we'll do an air quality plan under [Article] 23 in the UK'. Instead, the government should be closely monitored by the European Commission to ensure that they 'are not taking a free ride on other member states, who have taken the necessary investment to comply'.

The Court found that the UK government was in breach of its duty to comply with EU NO_2 limits in 16 cities and regions — including London, Manchester, Birmingham, and Glasgow. The European angle was perhaps the most perplexing for the law lords. 'The trouble is that the Commissioner has a rather small legal department,' one of them opined. 'I'm not sure if any of these people are necessarily lawyers. We don't get a huge amount from the Commission's legal attitude.'

'Perhaps we should refer the whole shooting match to the court and let them get on with it,' said another.

That particular court was the European Court of Justice, which sits in Luxembourg. And indeed, having found in ClientEarth's favour, the Supreme Court duly dispatched the case for that European body to give a ruling that would help clarify provisions in the EU Air Quality Directive.

It was time to head to Europe myself.

It took confidence to approach the Berlaymont building. Brussels houses the Parliament and the Council in walking distance of each other, and also this home of the European Commission. The Berlaymont is bowed like a vast wall, its mass of windows gridded with metal bars, like some high end prison that is intent on keeping its bureaucrats in and the public out. The EU flag of gold stars on a dark blue background covered what might have been doors. I kept walking, hoping for an open sesame moment, and it came. My approach triggered a panel to open and let me through to the building's bright, beech lined interior.

Elena Visnar Malinovska had decorated the walls of her office with the coloured drawings of her three children. The pictures were bright with sunshine and water and flowers. 'We are here, in the offices, close, like rabbits,' she explained, 'so I prefer to have a nice environment.'

Elena was a member of the cabinet of the EU Commissioner for the Environment, Janez Potočnik. Her brief included a numbingly wide array of environmental concerns, and among these was air quality. The European Commission was the chief enforcer of environmental directives, so was ClientEarth not usurping the Commissioner's authority? I had come to find out.

Was the work of such groups as ClientEarth important, or an annoyance?

Elena was clear. 'It's of crucial importance.' As the enforcer of EU law, the Commission could only act after the submission of a member state's annual report, which would be issued in September to cover the previous year. 'In member states, though, the public interest groups like ClientEarth could bring the authorities before courts. This could speed up the action on air quality tremendously. We feel sometimes like the elephant in a crystal store.'

That last phrase stems from Slovenia, where Elena was born and trained as a lawyer. She explained that it meant 'we have all the evidence, but we are just acting very slowly'.

So public interest environmental law groups brought speed of action, which was a net benefit for enforcement action in the EU. Clearly a sense of partnership was already established, as Elena spoke of ClientEarth's and the Commission's actions to enforce the Air Quality Directive in the EU as 'parallel exercises'. The Commission found no need to interfere when a national court appeared to be handling matters well. In Germany, when a citizen named Dieter Janecek demanded that Munich deliver an action plan to deal with air pollution near the ring road where he lived, Elena felt that the government actually responded to his challenge. That was not the case in the UK, where 'the court acknowledges the breach, but still has some question marks over some parts of the directive, and certainly does not give comfort to the complainant in the sense it does not say to the UK, "Now, you have to address the breach, and you have to adopt the appropriate measures." So this is where the highest court stopped short.'

Even the lower courts delivered what she felt to be 'very surprising decisions, in the sense that they weighed health against the economic impacts, and the possible costs of measures. I would say there is still some way to go for the British courts.'

And so in the wake of ClientEarth's Supreme Court victory, the Commission sent the UK government 'a letter of formal notice', the opening of a process of enforcement to which the UK had already sent a response. How was that response? 'It was murky,' Elena said. 'A bit arrogant.'

The Commission would move on to issuing a 'reasoned opinion', which would serve as the basis from which they would build a case. This was their 'parallel exercise', to run while the European Court of Justice considered the referral of ClientEarth's own case.

I wondered how she viewed the main environmental campaigning organisations.

'From the NGOs, we get mostly criticism. It's often like that. On air quality, after fighting very hard, we have been very angry.' The sense was that whatever they did would be attacked as too little and too late. That just played into the hands of the opponents, and even the political leadership inside the EU. If whatever they did was too little, she explained, then those leaders might wonder why they bothered at all, because they felt whatever they achieved could not satisfy the NGOs.

There was a plus side: 'They are always helpful in terms of briefing points, and they are involved in many processes much earlier than the cabinet or Commissioner is. But in terms of actual moral support, we felt there has been little.' She felt there was no recognition of how hard it was to pass green laws in the Commission, let alone the Parliament and Council.

'ClientEarth is somewhere in the middle.' She moved her hands apart as though holding a piece of string, and then slid her fingers to the halfway point. 'They have very good lawyers, but actually they also act as an NGO. I think we have got a lot of very good, well written analysis, and complaints supported by evidence. It's extremely useful that we get a legal environment outside the classical public institutions, like the Commission, the Council, and the Parliament. So I welcome this development very much; also they are not afraid of litigation. They defend the public interest, maybe better than groupings of neighbours, citizens, or associations.'

ClientEarth was currently the sole exemplar of a pan-European public interest law group, working across all areas of the environment. In what way could this sort of law group do more?

'I would like more of them. I would like them to do more legal assessment, looking for the legal particularities of different systems. This is where, on the Commission side, we lack staff. We have 40 colleagues, working on 28 member states. We cannot go after each case. What's certainly useful is if these public interest law groups actually take over certain complaints, and make the judicial systems more robust at the national level.'

Billboards still decorated buildings in Brussels urging citizens to vote in the European elections. They had done so, and the issue of the environment barely squeaked a mention in the electoral campaigns. As a consequence, many new MEPs for whom the green agenda was anathema were set to file into Parliament. I worried for the consequences. What was the best way of supporting the environment in this new political regime?

'There will be a very difficult time, I think, for the next Environment Commissioners, whoever they are, because the Parliament is not as green as it was before.' Elena was partisan by training, and so inclined to see that law could bring solutions. 'I am a lawyer, and I must say that regulation does deliver. Voluntary approaches in Europe do not always work. We don't have the Asian respect for the voluntary. Unless you have regulation, it's very difficult to call for respect.'

Tucked in against the grandeur of London's Selfridges department store, Duke Street is one of the dirtiest places on Earth. It looked clean enough, and indeed a diesel powered machine roared past me and sucked up litter from the pavement as I stood there. Taxis lined up in wait for shoppers, diesel engines purring.

A different cab drew up, this one a petrol hybrid. A young doctor climbed from the front seat, and three patients from the back. The patients had been through early morning health checks at the Royal Brompton Hospital, and the hybrid cab continued the attempt to keep them in as healthy an environment as possible. Now that was over. For two hours, they would march up and down a hazard zone while any deterioration in their health was measured.

The trio of patients blew air through a cardboard tube, on and on till their lungs were emptied. That job done, in just a few steps they were among the crowds on Oxford Street.

It was a sunny day. London was showing off two of its icons: a stream of bright red double decker buses and a flow of black taxicabs. Other traffic is kept away, which makes Oxford Street pretty special for medical researchers. The high stores on either side turn this London thoroughfare into a canyon, one pumped with regular discharges of diesel exhaust.

Michael and Ashif had heart attacks a while back. This bout of air pollution is likely to stiffen their arteries and increase cardiovascular risk. Gorana has COPD, chronic obstructive pulmonary disease. Their previous outing, in this controlled experiment run by King's College London, was a walk in the relatively diesel free air of Hyde Park. Gorana would never come to Oxford Street by choice. Now the experiment had brought her here, she tended to linger by the displays in shop windows whenever she could.

In London's 'pea soup' fogs of the early 20th century, Gorana would not have seen as far as the shop window's glass. The Clean Air Act of 1956 condemned those fogs to history. The new breed of air pollution is more insidious: it streams invisibly from diesel exhaust as the toxic gas nitrogen dioxide, which boosts levels of ground level ozone; or it is spewed as fine particulate matter.

I trailed in the wake of the experimental team, and caught up when the three patients paused to breathe into their tubes once again. Their breaths were shorter, and the difference measured their deterioration in lung function. These two hours of tramping up and down Oxford Street felt like a killer, even to me. My glands swelled and my lungs burned. For years I have sought out back routes, so as to avoid walking alongside heavy traffic. It was a simple act to protect my own health. I never considered ways of helping others: a King's College survey of 2010 estimated 9,500 annual deaths from air pollution occur here in London. This experiment was focused on the extra burden of nitrogen emissions.

Since I was stuck on Oxford Street, this is what was happening to me. I was walking through an invisible cloud of toxics. They were charging deep into my lungs and had likely entered my bloodstream.

Forty healthy volunteers were tested alongside 40 who had COPD. Rudy Sinharay, the doctor who led us up and down between Oxford Circus and Bond Street with remarkable speed and enthusiasm, would later announce: 'These findings show that airways obstruction and a stiffening of the arteries occurred in both the healthy volunteers and people with lung disease — even after limited exposure to diesel pollution.'[16]

The patients that performed the experiment stayed remarkably cheerful, and zipped back to the hospital to complete the day's tests. I went into Selfridges for a cup of tea. When I spoke to order it, no voice came out of my mouth. The pollution had snatched it away for a while. I managed a hoarse whisper, and waited to recover.

King's College sited a monitoring station by an entrance to Selfridges. It checked NO_2 levels on Oxford Street day and night. The EU's limit was 40 micrograms per cubic metre of air (40 mcg/m^3). The Oxford Street average more than trebled that at 135 mcg/m^3, and this was for periods which included the less poisoned night-times. Shortly after my visit, a daytime reading peaked at 463 mcg/m^3. Findings were released to *The Sunday Times* in July 2014.[17] 'Oxford Street worst in the world for diesel pollution' ran their headline, and the reputation belched into further headlines across the world.

Dinah Rose's statistic of 29,000 annual UK deaths from air pollution related to people who die from breathing particulates. Nitrogen dioxide has its own lethal tally to add to the numbers. Professor Frank Kelly, who chaired the government's Committee on the Medical Effect of Air Pollutants, announced that nitrogen dioxide 'would increase air pollution's contribution to the total death rate from 5–9 percent across the UK to 10–18 percent'. Add together the effects of $PM_{2.5}$ and NO_2, and the estimate was that diesel cars were killing 60,000 people a year in the inner cities of Britain.[18]

The air quality case was passed from London's Supreme Court to the European Court of Justice. Teams of lawyers travelled to Luxembourg. Maybe those lawyers spent ten minutes drinking a cup of coffee the following morning. In those ten minutes, UK air pollution killed someone new.

I place these lawyers in the context of time passing because in many ways that is what the whole case was now about. Nobody was any longer arguing the need for the UK government to bring in a plan to end this death by pollution. That was agreed. The disagreement was over when it should happen. One side were proponents for urgency and the other for delay. A scrap of everyday language was once more the focus of legal debate: the phrase 'as soon as possible'.

As a proponent of delay, the UK government was setting new records. Lawyers from the European Commission told the European judges that they were considering 'perhaps the longest-running infringement of EU law in history'. It was 'a matter of life and death'.[19]

The UK missed an original 2010 deadline, and had no intention of meeting the next of January 1, 2015. The new date they held up for compliance was 2025. In correspondence with the EU, the UK admitted that hopes for legal NO_2 levels in London, Leeds, and Birmingham had slipped back to 2030.

'Legally privileged information!' the UK's lawyers declared when the information was submitted to the court.

ClientEarth's lawyers duly pointed out that the figures had been posted on Defra's (the Department of Energy, Food, and Rural Affairs) own website just the night before.

'There were lots of red faces, confusion, and embarrassment in Defra's corner,' the ClientEarth lawyer Alan Andrews remembered. 'I don't know whether it was cock up or conspiracy, but it put the UK in a very bad light.'[20]

The European court held that the UK government was indeed obliged to mitigate air pollution, and was wholly adrift of all procedures to fit such compliance to the given deadlines. Furthermore, it must produce a plan to keep the period in which nitrogen dioxide pollution was breaking legal limits 'as short as possible'.[21]

So that time phrase, still exceedingly light on definition, was duly wrapped up and sent back to the UK Supreme Court.

Let's stay in mainland Europe for a while. After working with the NATO/ EU Peacekeeping force in Bosnia, Małgorzata Smolak (known as Gosia) returned home to Kraków.

Kraków draws in nine million tourists every year. Its medieval square is magnificent, and a 14th century castle sits proud above a bend in the Vistula River. You expect such a city to top league tables. And it did. Kraków came eighth out of 575 cities worldwide — for levels of $PM_{2.5}$ in the air, those clouds of tiny particulate matter that provoke asthma and cancer.[22] People cough all winter and they lose their voices, and it's normal. 'That's how life is in Kraków' was the general consensus. It's how it's always been.

Gosia became a lawyer with ClientEarth. She wanted her fellow citizens to breathe better air. Even more than that, she wanted them to take on the fight to win clean air. They would feel more powerful with every breath.

Half Kraków's pollution comes from coal fired power plants and traffic emissions. The other half comes from domestic stoves. Just one in ten households uses those stoves.

The solution seemed easy: phase out those stoves and ban them.

Getting Crakovians to accept a ban was like breaking a horse. Communist authorities imposed draconian restrictions, so anything that now threatened hard won liberty was resisted. The trick was not to impose a solution on the citizens, but to have them claim both the need and the solution for themselves. Kraków Smog Alert (*Krakowski Alarm Smogowy*) formed an alliance with ClientEarth Polska and the Polish Green Network (*Polska Zielona Sieć*, itself an alliance of green groups). The goal of cleaning Kraków's air became a genuinely popular one. Newspapers made it a headline story, and Gosia joined 2,000 citizens as they marched through the streets.

How do you convert a popular campaign into a practical outcome? ClientEarth used Poland's Environmental Protection Law Article 96, which allowed for a ban of fossil fuels for environmental protection purposes. ClientEarth's legal briefs to Małopolska Council (which governs the province containing Kraków) gave them the courage to move forward with a ban on the burning of solid fuel in the city's stoves. The council chamber erupted in cheers and hugs as the result was announced.

It's fun to celebrate, and wise to remain on edge.

Two private citizens challenged the verdict. An appeal to the Supreme Administrative Court brought a declaration that the powers of Article 96 had been stretched too far.

Anticipating the judgement, ClientEarth lawyers worked on the language of Article 96. They sought to make the requirements more precise, and to give more exact powers to regional and local authorities. With colleagues from Kraków Smog Alert, they lobbied the government. In October 2015, the President of Poland, Andrzej Duda, himself a lawyer and a native of Kraków, signed the revised Article 96 into being. Kraków's ban on domestic stoves could be put into effect once again.

The new ban, should it survive any further challenges, will enter into force in 2019. Gosia's fellow citizens hope to breathe clean air soon, and, with every breath, they will feel their own power to achieve change.

Before taking a break from Poland, let's follow lawyers up into the mountains. What they achieve there will filter back to Kraków and other Polish cities.

Zdzisław Kuczma lived in the Silesian town of Rybnik, best known for its coalmines and power plants. For a break, he headed up to the mountain resort of Wisła. It has world class ski facilities in the winter and mountain valley hikes in summer. You might hope for clean air up there.

Polish law allows a clean air tax. Municipalities can impose tax on tourists to pay for touristic infrastructure and environmental standards. The tax might be as small as two zlotys a day, but that can add up to millions in income for the town. Mr Kuczma paid his hotel bill, including the clean air tax, but smelt something odd. The air of Wisła seemed as filthy as the coal fug he got to breathe back home.

ClientEarth Polska's lawyers studied the air quality data. Something was indeed wrong. These mountain health spas heated themselves with domestic stoves and maintained some of the most polluted air in the country. They were collecting taxes to clean the air, without bothering to clean the air at all. The lawyers joined Mr Kuczma in lodging an official challenge to the taxes.

In November 2015, the Regional Administrative Court in Gliwice passed judgement. Wisła's collection of local taxes for clean air was illegal. Wisła was not alone in imposing such charges. Over 900 towns and cities have been collecting the taxes, including Kraków with its filthy air. The day following the Gliwice court decision, Kraków announced its collection of this tax would stop.

From now on, to collect the tax, they will have to clean up the air. This will provide an economic incentive on top of their existing legal obligation.

In the main courtroom of the UK's Supreme Court, rabbits and dragons are carved into the ends of dark wooden benches. Winged angels are fixed among the wood of the rafters. The courtroom is placed in the context of a mythical natural world, but was being asked to save the real one. The air quality case had been batted back to it from Europe, and now there was nowhere else for it to go. The law lords were required to settle it once and for all … or at least, for now.

Two questions excited the law lords as they began. The European Commission was chasing down 17 other member states when it came to particulate matter, but only the UK in the case of nitrogen dioxide. 'Why is the Commission singling us out?' asked Lord Carnwath, the lord with the keenest environmental pedigree. He looked down at papers for an answer, his mouth pursed in puzzlement. 'There's no indication of why it's only singling out the UK.'

'It was James Thornton,' I could have explained, if comment were acceptable from the public benches. First, James had worked to change the cost rules and so ensured that a case could be brought into the UK courts, and then he had chosen this one to go all the way. But James was not singling out the UK. He was just starting there, with the intention of rolling out similar cases across Europe. He sat at the curve of tables behind his newly appointed barrister Ben Jaffey, to keep an eye on the action.

And then, I might have added, as I had learned from the Cabinet Member responsible for air, the Commission filed their complaint because of this Supreme Court's lack of assertion in their first hearing. But I stayed silent, and the lords' next question duly arose: the 'as short as possible' one.

'Would you accept that what is meant by "as short as possible" is not necessarily concerned with absolutes,' Lord Sumption probed. His shock of white hair flew behind him, like he was at the wheel of an open top sportster. 'It is what is realistically possible,' he suggested as an alternative.

Ben Jaffey blinked rapidly in the face of the questions. An absolute would mean removing all traffic from the streets at once, Sumption continued. 'So you could ban all vehicles from the streets?'

'Yes,' the barrister declared, for his reading of the law said that such questions as affordability should not affect prompt action.

'I hope you have a fallback position,' responded Lord Carnwath.

I laughed. Laughter was allowed, apparently.

Kassie Smith QC was fighting the government's corner. Hers was a tale of a government rendered incapable of submitting a deadline under one article of the law because it was already in breach of an earlier one. Since this bureaucratic tangle had rendered them unable to act, they should escape punishment. Lord Sumption briefly told her why her argument would not wash: 'It's essentially a "not being allowed to take advantage of your own wrong" argument.'

Even as the QC told of the government's efforts at a plan, she worked to turn it into a justification of delay: 'The air quality plans are going to be complex and detailed,' she assured the judges, 'but these are also issues that take a long time to address.'

She paused her pleading for lunch, and resumed it afterwards. From the benches, her chances did not look good. Lord Sumption only managed to control his yawns by letting his head tilt back against its rest, his eyes shut, his mouth open. Lord Neuberger, presiding, held his biro in his mouth like a long cigar. Clenched between his teeth, it swivelled left to right and back again, as though tracking something that was worth writing down.

Government was 'in purdah', mid-election and so devoid of executive powers, but the QC did allow that their action plan could be ready in five months. As the information leaked from her mouth, the lords' next move became clear. The government had admitted its ability to deliver a plan. To ensure some parity between what a government says and what a government does, all the court needed to do was impose a mandatory order. Allow for the election and slot things within the calendar year, and the timeframe became obvious. It revealed itself in a rapidly returned judgement two weeks later. 'The Supreme Court unanimously orders that the government must submit new air quality plans to the European Commission no later than 31 December 2015.'[23]

Lord Carnwath had earlier pointed out the court's limited scope of action, since imposing a fine would have required the government to pay itself. The mandatory order, imposing a strict deadline for action on the government, had been ClientEarth's principal demand, and so in this way the judgement was a total victory. And effects were immediate. The committee tasked with investigating airport expansion plans in London delayed its report in order to account for air quality issues, since the law was now requiring a compliance with air quality directives that had been routinely ignored before.

What courts can do for you is limited by what you ask of them. The Supreme Court was newly formed, freed of its governmental shackles, but weighted by centuries of deference. James had fought to demand more of the Court — specifically some fixed term to 'as short as possible'. A precedent had already been set by the EU when it limited delays to previous deadlines to five years. The barrister argued that if you ask too much of a court, then it gives you nothing.

The mandatory order demanded 'urgency' of the government, but they were wedded to delay. How would they understand 'as soon as possible'? James required his barrister to get the Supreme Court to let the parties go back to the High Court to clarify the order. That provision was duly granted.

James was trained in the US legal system. Back in 1803, the American Supreme Court was relatively new. A case decided that year, Marbury

v. Madison, is a foundation stone of rule of law for every American lawyer. It established that the Supreme Court, not another branch of government, has the ultimate power to decide what the law is, and to enforce it.

In the UK in 2015, the Supreme Court was a new institution, which grew out of the House of Lords. It was still defining its powers, especially in relation to the European Commission and the Human Rights Act. In ClientEarth v. Defra, the Supreme Court ordered the government to comply with its legal duty. It also assigned the courts to supervise compliance with the court's order. For James, ClientEarth v. Defra has some of the quality of Marbury v. Madison. Which is why the UK government's response appeared as an attack on democracy.[24]

'The government stood up in the Supreme Court in 2015 and admitted they were out of compliance from 2010, saying they could not comply before 2025. After the Supreme Court ordered compliance, they had the defiance to produce a plan for 2025.

'Forget that this is environmental law. It could be contract law or family law, company or any other kind of law,' is his take on things. 'The Parliament transposed the Air Quality Directive into UK law, and the government must obey. Otherwise, we don't have democracy under the rule of law, we have rule by fiat and diktat of whoever is in power. In the dialogue of power the government is engaging in, democracy requires us to support the courts by giving them the opportunity to correct the government.'

In March 2016, the UK Environment Secretary, the Scottish and Welsh ministers, the Mayor of London, and the Department for Transport were served with papers that were lodged at the High Court in London. Lawyers on all sides duly went to work. The Head of Air Quality at Defra alerted the minister responsible for air pollution to her findings. 'The implementation of clean air zones was vital,' she assured him. 'These would bring forward compliance [with EU NO_2 limits] by directly removing the dirtiest vehicles from hotspot areas and by encouraging people to swap polluting vehicles for less polluting ones.'

Her clear analysis came with a rider: 'There is resistance from other departments about clean air zones including cars.'

Documents presented during the two day October court hearing revealed the main villain, the chief resister, to be the Treasury. The department reduced Defra's plans for 16 clean air zones down to just five outside London, for fear of scuppering schemes to boost the economies of northern cities. To avoid political backlash from angered motorists, they also blocked an increase in charges for driving in city centres.[25]

News media ringed the Court for the judge's ruling two weeks later. 'I reject any suggestion that the state can have any regard to cost,' Justice Garnham declared, and noted as 'remarkable' that ministers based their overoptimistic pollution modelling on lab tests they knew to be flawed, rather than actual emissions on the road. He granted ClientEarth their second court victory over the UK government in 18 months.

That same afternoon, the Prime Minister rose to her feet in the House of Commons. 'We now recognise that Defra has to look at the judgement made by the courts, and we now have to look again at the proposals we will bring forward,' she admitted. 'Nobody in the House doubts the importance of air quality.'[26]

The victory was intended to empower citizens to use law to defend the environment. A quiet victory would not do the job. In the 24 hours after the High Court judgement, the verdict triggered 1,400 media hits around the globe, including on all the main UK national news bulletins and the front pages of *The Times* and *The Daily Telegraph*, who both featured leader articles. And almost 30 MPs from across the political divide rose to their feet in Parliament to demand urgent action on improving air quality.

'The political tide has shifted, and more and more people realise that pollution really is something that needs to be tackled,' Tim Reid, Director of Communications at ClientEarth, reflected. 'That's been done through a joint effort between the legal teams and the communications team — working strategically, hand in hand, every step of the way. By raising this as an issue that everyone should be worried about, by keeping up the pressure, and by highlighting the health impacts every day, the media is at long last seeing the invisible killer as one which they need to focus on as well.'

'Invisible killer' is an emotive phrase, but language in fact falls way short of the reality. The 20th century saw the coinage of the term 'genocide'. The 21st century needs something similar for air pollution. It kills more than half a million Europeans each and every year. The killer's very invisibility means that we all feel we can get away with unleashing it, and so we need education and restraint. In governmental terms, it is a slaughter of a country's own citizens through wilful neglect born of economic priorities and political exigencies.[27] Eighty per cent of these premature deaths are from stroke or heart attack, with lung disease and lung cancer making up the bulk of the rest. Beyond that, fresh evidence links air pollution to new onset type 2 diabetes in adults, obesity, and dementia. It stunts neural development in children and blocks their lungs from growing to full capacity and so leaves them prone to asthma. The living landscape they grow up in is damaged by air pollution, and their built environment crumbles. The European Commission has estimated the total annual health related external costs of air pollution to be in the range of €330–940 billion.[28]

Such data evaporates when the government is lobbied by powerful commercial interests to grant approval for schemes that have high environmental impact. 'You will be aware of the recent judgment in the ClientEarth judicial review of the government's Air Quality Plan,' Chris Grayling, Minister for Transport, wrote to the government's Environmental Audit Committee as they considered the decision to build a third runway at Heathrow Airport. 'We are carefully considering what this means for the air capacity programme.'[29]

The judicial review found the Air Quality Plan to be 'woefully inadequate', particularly with reference to new evidence on real road emissions. 'We will want to hear from the minister how the Government can meet air quality standards,' the Audit Committee's Chair responded, 'given what we now know about real-world emissions, which are higher than used in the Government's business case [for a third runway].'[30]

Lawyers hold politicians to account when they conspire to kill us. It is a lonely business, and expensive for a charity. Relax their guard for a moment, and forces that profit from environmental devastation seize their chance. Hopefully media attention will help the rest of us awaken.

In the UK, people become vocal when the National Health Service comes under attack. While environmental concerns made token appearances in the UK's 2016 EU referendum, a promise to divert payments away from the EU and into the hospital system boosted the Brexit vote. A dash of extra intelligence would show that 50,580 annual UK deaths from $PM_{2.5}$, NO_2, and O_3 alone place an intolerable burden on the country's health system. We must add to that the 'much larger' proportion of the population who need treatment for the less severe effects of air pollution (i.e. they are sick yet survive).[31] ClientEarth's lawyers are now enforcing the Ambient Air Quality Directive in courts across Europe. In the UK, they now also have to work to ensure a new Clean Air Act is robust and informed by the best global standards. This is citizens' action at work. The least we citizens can do is appreciate its primacy in our lives.

The sheriff comes to town

James Thornton

As a young lawyer, I specialised in enforcement for a number of years, and brought so many cases that it embarrassed the Ronald Reagan government into enforcing again. This experience taught me that citizens with limited resources could still enforce the law. It also taught me how important enforcement is.

If you knew that the government never audited tax returns or prosecuted people for non-payment, would you be tempted to fiddle your tax returns? If you were a company and knew that to reduce pollution cost money, and money is your only grammar, would you be tempted to pollute?

Enforcement is critical at every level. If you have a dog, you know you have to train it to behave, with negative feedback for bad behaviour. The same is true of governments.

In the EU, scholars have long talked about an 'enforcement deficit'. Good laws are passed, but not well enforced, or not enforced at all. In Italy, there is a proverb that has emerged since the EU was formed. It goes like this: one goes to Brussels to make the laws; one comes home to find a way around them.

The Italians may give voice to what northern Europeans keep to themselves. While compliance is better in northern countries, it needs improving everywhere. Comparing the USA to the EU, compliance is better in the US.

The image I have of EU politicians passing good laws is of people who want to take credit for a job well done. They pass a good law. They get good press, citizens feel good. Everyone knows it won't be enforced. Sometimes the law is written in such a way as to make enforcement difficult. Sometimes the relevant enforcer does not want to do the job.

Knowing how important enforcement is, and discovering the enforcement gap in the EU, I decided before opening ClientEarth that we would have to be in the enforcement business. My goal was dual: to create more respect for the rule of law among companies and countries, and to make it clear that when ClientEarth spoke, you should listen. As the Polish Treasury Minister was later to say, 'ClientEarth is a bad enemy to have.' Words of high praise if you want to get the job done.

I had a strategic choice to make. What law would I enforce, and where? The UK offered good access to the courts, and while the rules on costs were far from perfect, we had moved them in the right direction.

The next question was which law to enforce. I had learned from my work in the US that I needed a law that had extremely clear behavioural requirements. Ideally there should also be deadlines.

I read a lot of EU law, and conferred with experts. In the end, the Air Quality Directive was ideal. It had numerical values to limit pollution, it made the government the responsible party, and it had clear time limits.

Not only that, but noncompliance with this law has caused a public health emergency throughout the EU. Some 600,000 people a year die of air pollution in the EU,[1] primarily due to Europeans' love affair with diesel.

So there was deadly pollution, enforceable limits, deadlines, and responsible parties. We started in the UK and won, ultimately in the Supreme Court, which ordered the government to clean up the air. The government was violating the court order, so we went back to enforce it.

Meanwhile, we are rolling out a clean air enforcement campaign throughout Europe, with ten cases now filed in Germany, and more to come in other countries. Our strategy was to demonstrate that citizens could enforce the law, and then do it broadly. We began the European leg of our enforcement work in Germany, but air pollution is a serious threat to public health right across the continent. In many EU countries, only NGOs domiciled in the country can be a plaintiff, so ClientEarth cannot be the plaintiff itself. We follow the model I used back in Chesapeake Bay and find an in-country environmental group to work with. In Germany, we work with DUH, the national branch of Friends of the Earth. They serve as plaintiffs. Their communications and outreach benefit the endeavour, while we map EU strategy, and hire the lawyer to bring the cases. Belgium, the Czech Republic, and other countries will be further venues for this work.

This EU air campaign is demonstrating that governments need to comply. It will deliver cleaner air and protect the lives of citizens. It will provide climate benefits as well, since the black carbon in diesel absorbs sunlight and raises temperatures. It should close down some dirty coal power plants, giving further climate benefits.

And, to end where I began, it shows that citizens with relatively little resource can make the law effective by enforcing it. Making this one law work should also send the message to start complying with environmental law more broadly. Citizens are watching and can act.

6

Leaving Plenty More Fish in the Sea

This is how the future of our seas looked in 2006. A worldwide group of leading ecologists undertook 'the most comprehensive look yet at the human impact of declining marine biodiversity'. Their paper in *Science* projected the collapse of all commercial fish and seafood species by 2048.[1]

'It's a gloomy picture,' admitted its lead author, Boris Worm.[2]

In 2007, the Food and Agricultural Organization of the UN reported that global fish stocks were 75 per cent overexploited, and noted 'the maximum wild capture fisheries potential from the world's seas has probably been reached', which called for 'a more cautious and closely controlled development and management of world fisheries'.[3]

Recent environmental groups formed around single issues: focused on oceans, maybe, or rainforests or climate change. Because all ecological problems are connected, James was determined to bring legal expertise to all bases. That would be a broad scope to take on, even for the United Nations. For a fledgling law group, it clearly needed a tight strategy.

James's strategy came in six parts, posed as six questions. Here's the opener:

What's the most important problem?[4]

The loss of all fish from the sea was a most important problem, and overfishing was its chief cause.

The line tweaked and then started spinning out fast. The other anglers laid their rods in the skiff and joined in. However big this fish, it was mighty powerful. They helped to reel it in.

The halibut appeared, an 80 pounder. Two bulging eyes and a whiskered snout appeared as well. While the anglers pulled from one end, a sea lion tugged at the fish's tail from the other. It was a tug of war. The tail ripped, and the fish landed in the boat.

Sixteen wealthy people raised halibut to their lips. So did James and I. We were courtesy passengers on *Mist Cove*, one of two ecotourism boats run by the McIntosh Foundation in Alaska. The stern was transformed into a restaurant with a shifting view over Admiralty Island.

Our visit was in September 2010, on the last boat of the season. Back in Europe, a tiny team at ClientEarth was preparing to tackle reform of European fisheries. Those seas range the Mediterranean and around the eastern Atlantic to the Irish, North, and Baltic seas, and contain multiple species. Alaskan waters are broad and simple: skiffs would head for one stretch, pretty sure of finding halibut, or another from which to haul out salmon. The journey helped me set a few key concepts in place, before seeing how lawyers approached Europe's regulatory mesh.

In the vast sheds of an Alaskan salmon hatchery, vats brimmed with salmon eggs. Fry crowded the outdoor tanks, where kingfishers dive bombed them from trees. The fry are released into the sea to travel and grow, and return to their birth waters to spawn. Trawlers crowded the basin outside the hatchery and hauled the salmon aboard in swollen nets.

One single trim waterfall leads up through concrete from the open water where the salmon crowd. That waterfall is the salmon's only chance to reach the higher breeding pools. A trap at the top determines whether they will get through or not. Humans are so efficient at harvesting the eggs and birthing the fry, they seldom need to open the trap. It was closed today. Fish hurled themselves against it, and fell back.

We stood on a concrete apron and looked down.

Two brown bears strolled from the tree cover and pawed salmon aside to the concrete slope. With so much food, the bears had become the epicures we learned of earlier. They sliced the fish open for their hearts, nibbled and sucked, and then left the bulk of each fish to rot.

They didn't, however, swipe every salmon onto the land just because they could. Individual salmon blood flowed down the concrete, but this was no killing spree. Bears don't do that.

That's probably why bears don't rule the world.

A hundred and fifty years ago, the writer George Perkins Marsh noted how Timothy Dwight, the President of Yale College, travelled to the shores of Long Island Sound in the 1820s, where he discovered that 'white fish, a species of herring too bony to be easily eaten', were being used as manure. 'Ten thousand are employed as a dressing for an acre, and a single net has sometimes taken 200,000 in a day.'

Marsh was not starry eyed about fish. 'True fishes are extremely voracious, and almost every tribe devours unsparingly the feebler species, and even the spawn and young of its own.' He was not starry eyed about humans either, and linked them into this predator–prey chain of relationship:

> The enormous destruction of the pike, the trout family, and other ravenous fish, as well as of the fishing birds, the seal, and the otter, by man, would naturally have occasioned a great increase in the weaker and more defenceless fish on which they feed, had he not been as hostile to them also as to their persecutors.[5]

Marsh chose freshwater fish as his indicators. When inland waters were fished out, humans switched focus to the more dangerous seas. Marsh published his *Man and Nature* in 1864, five years after Darwin's *On The Origin of Species*. The reach of its research was wide, covering the globe and the historical record, and its impact was profound. Between them,

Darwin and Marsh 'put paid to traditional faith in a designed nature and preordained harmony between humanity and the rest of creation'.[6]

The removal of that belief in a harmonious whole revealed a fractured world. Humans had set themselves apart. The natural world was their bounty, and they assumed rights over it. Laws decreed how men (for women had no such rights as yet) could divide the spoils.

It would take public interest lawyers 100 years to come on the scene and point out that it wasn't just a matter of squabbling over the spoils. The natural world had rights as well. By then, of course, the planet was in an almighty mess.

In 2000, scientists Paul J. Crutzen and Eugene F. Stoermer published an article that established the term 'anthropocene' to describe our new epoch. The nub of their argument was restated a decade later as 'humankind has become a geological force in its own right', a species 'so large and active that it now rivals some of the great forces of Nature in its impact on the functioning of the Earth system'.[7]

Crutzen and Stoermer's historic article ends with this statement: 'An exciting, but also difficult and daunting task lies ahead of the global research and engineering community to guide mankind towards global, sustainable environmental management.'[8]

In seeking a solution to what is ultimately a matter of rights, they called on the research and engineering community. Even they forgot to call in the lawyers.

Any lawyer for halibut might start with establishing one right: Let the fish breed.

A female halibut reaches 50 per cent maturity in 12 years. Her ova will ripen in batches, and she will release them whenever they are ready. Once she has spawned for the first time, she will continue to spawn annually, if she is allowed. Her spawning season is November to March each year.[9]

November 16, 1924 rides high in the annals of Halibut History. The International Pacific Halibut Commission brought in the first three month winter moratorium on fishing the species. It made obvious sense. Let fish breed and there will be more to catch. Besides, who really wants to fish Alaskan waters in the winter?

Regulators hooked onto that closure period as the way to manage stocks. The period kept expanding, until in the early 1990s the complete annual catch was compacted into a 'matter of hours with subsequent landings unloaded on overwhelmed processors'.[10] Quality was poor, prices were low, and all fish had to be frozen. Beyond that, for the fishers, those few hours became an infernal chase.

Halibut can live to 55 years, and grow to two metres long and hundreds of kilograms. Drop a couple of 36 kilogram giants in your boat and you'll know it. In the 1991/1992 chase for halibut, boats became overloaded, and nine men were killed. Individual transferrable quotas (ITQs) were brought in as the solution in 1995, and the season duly expanded from two days to more than 200. A total catch was set, rights were distributed to owners and associations, and these new owners were allotted a percentage of the total. That allowable catch has been described as 'equal to the efficient catch for the fishery'.[11]

Opinions of the quota system depended on who got the quota. Those who went without were not fond. The higher the quota, the greater the degree of satisfaction. It is said to have worked 'well in economic terms, ending the race to fish and stabilizing the flow of fish to consumers'.[12]

A crunch word there is 'economic'. Global studies in the transferable quota system are acknowledged to be limited, and to largely focus on economic rather than 'real world' impacts.[13] Those economic impacts include the fact that 'the free-market form of catch shares adopted in the United States, Canada, New Zealand, and Iceland has tended to put fishing rights in the hands of a wealthy few, often absentee, quota holders'.[14]

Alaska maintains the proviso that quota holders must be on board their boats, and not behind desks where they count the rents their quotas bring in. Core to the success of fisheries management is such hands on engagement by those who know the seas on that daily basis.

In Alaska, halibut exists in a single species fishery. The North Sea, in contrast, is a great place for fish that like mixed company. It's a shallow basin, often less than 40 metres deep in places, and so it heats up rapidly. The sea managed a spike of a 0.9°C increase in 1989 alone. Between 1977 and 2001, two thirds of the fish we choose to eat had already shifted location. The species that grow fastest nip up there to revel in warmer waters. They hang out with the larger cold water species who migrate north more slowly. It takes high technology for humans to come along and pluck one species out of the crowd. Europe's fishing fleet is pretty advanced in that way and, to use the appropriate industrial language, 'have demonstrated capacity to maintain high catch per unit effort even as stocks decline'.[15]

The language of fish conservation tends to such terms as 'stocks'. Fish are known as a CPR — a 'common pool resource' — and their fates are determined by RFMOs — Regional Fisheries Management Organisations. The focus is on maintaining supply to the top of the food chain, that top being humans, while simultaneously preserving the industry. Professors Samuel Barkin and Elizabeth DeSombre explain how:

> Global fisheries suffer from a problem of regulatory capture, where the entity entrusted with overseeing the public interest instead acts in the interests of industry, as though 'captured' by it … It is also arguably the only resource-extraction industry for which as a general rule there is no regulatory oversight by separate environmental regulators who are bureaucratically distinct from the industry regulator.[16]

Thus global warming meets geological factors in the North Sea, fish respond to the temperature rise, a sophisticated fleet knows how to chase each species down, and the industry has an overwhelming voice in regulating how many fish can be pulled from the sea.

Europe's Common Fisheries Policy (CFP) was first formulated as part of the Treaty of Rome in 1956. It evolved through the decades into an agreement that is revised every ten years. The CFP is supposed to provide for a healthy fishing industry in Europe, with well stocked seas. In 2012, the CFP was up for reform.

The fishing industry lined up its lobbyists to influence proceedings. On the other side, a group of 193 NGOs formed a coalition called Ocean2012, which coordinated actions with other large NGOs, such as WWF, Greenpeace, and BirdLife.[17]

It was a crowded field. And while the whole process was one of law reform, those charities offering advice were strikingly devoid of dedicated legal expertise. While still a fledgling, as the only pan-European environmental legal group, ClientEarth was well positioned for impact.

James got some seed funding from the Oak Foundation in Geneva, and used this to assign Sandy Luk to the task. James's brief to her was wide. Do a broad examination of the state of fisheries, to take in the science and the applicable laws. And then design a clear approach, unfettered by previous methods, through which the law could be refashioned to correct what was going most grievously wrong.

The blue sky process began in a summer garden, when Sandy quizzed a scientist from the Marine Conservation Society, Melissa Pritchard, about the main issues from a scientific viewpoint. Overfishing floated to the top of the discussion, a core difficulty being that different fish species shared the same patches of sea. Cod could not elect to enter nets while haddock chose to swim on by; they were both hauled out of the water together. Then came trouble. If the fishers only had a quota to catch cod, the haddock did not belong on the boat. They had to be chucked overboard.

What emerged from the blue sky period was a plan for a fishing credits system. Essentially, this was a quota based system that did not just take into account the individual fish stock status, but included such elements as each species' position in the food web and its importance as a predator–prey species. It allowed fishers to deal with mixed catches but also took account of the environmental impact of fishing. This is an ecosystem approach that 'relies on the need for sound science, adaptation to changing conditions, partnerships with diverse stakeholders and organizations, and a long-term commitment to the welfare of both ecosystem and human societies'. In the mixed fisheries of European waters, such concerns are especially vital, since 'fishing activities usually affect other components of the ecosystem by catching non-targeted species, damaging habitats, interrupting the food chain, or reducing biodiversity'.[18]

In itself, this credits system was too radical, too complex as to its modelling, to swing a body of support in its favour. As the first statement in a theory, however, it established one core goal by which other proposed measures might be evaluated. The need could be stated in basic terms: set fishing levels to allow fish stocks to reproduce and survive indefinitely.

It sounds almost too simple to be worth stating. However, a main thrust of the European Commission's proposals for fisheries reform had a different ambition. They aimed to issue individual transferrable quotas (ITQs), under the theory that if you give people a property right in a resource, there will be much more stewardship; their long term interest should mean they work not to overexploit the resource. Those who were not able to survive on the fish stock they had could sell their right to it.

The Environmental Defense Fund (EDF) sent a high level posse to the capitals of Europe to support this focus on transferrable quotas. Set up in the USA in 1967 and now with over a million members, in 1991 *The Economist* called EDF 'America's most economically literate green campaigners'.[19] Positioned as the motto on their website, that quote clarifies their thinking. Their website's fisheries page remains unequivocal: 'EDF recommends designing rights-based management systems'.[20]

We have seen how this system worked in Alaska. As ClientEarth's lawyers entered the European reform process, the European Commission proposed to roll a similar system out across the continent. It took me a while to understand how effective techniques in one area might not apply to another.

In 1982, the United Nations Convention on the Law of the Sea gave countries control of waters up to 200 nautical miles (370 kilometres) from their shore, and so they were able to begin to disperse property rights. In Europe, EU laws mean the fish stocks in European waters are shared. John Lynham, co-author of an influential paper that recognised advantages in the transferrable quota system,[21] drew this conclusion from visiting fisheries around the world: 'The people who live close to the resource and are actually harvesting it are the ones who care about it the most.'[22] The large fleets have the capacity to chase down target species, but their degree of sensitivity towards local ecology many hundreds of kilometres from their home harbours is questionable. 'Implementing ITQs on a large scale can easily cause a large decline in small-scale fisheries (SSFs), or even their disappearance,' writes Adam Soliman about the individual transferable quotas:

> ITQs are likely to have this effect because their tradeability rewards economic efficiency, and in fishing as in most types of productive activity, greater efficiency tends to result from larger size. Small fishers, because their unit costs are almost inevitably higher than those of large fishers, will find it more profitable to sell their quotas to large fishers than to continue fishing.[23]

These transferable quotas 'can be a useful management tool', according to James, 'but only in a regime where you set the appropriate fishing level. They are not the silver bullet to solve all problems.'

ClientEarth's lawyers took on the mission of deflecting the discourse away from such quotas, so the focus could switch to ensuring that the fish catch was of sustainable levels. Briefings were sent to all interested parties. Ulrike Rodust, a German MEP and member of the German Social Democratic Party, was tasked with the role of 'rapporteur' to the Fisheries Committee. 'I regularly met representatives of ClientEarth during the process,' she recalled, 'as they offered strong and detailed legal expertise to all important aspects of the Common Fisheries Policy, especially to the issue of individual transferable fishing quotas/concessions.'

The winning argument against an EU wide enforcement of a transferrable quota system was ultimately ClientEarth's legal one. As one lawyer mentioned, 'Briefings are about unintended consequences of kneejerk reactions.' In their briefings, the lawyers pointed out that bringing such a rights based system contravened EU treaty obligations. Transferrable quotas were duly dropped.

I was due a break. If I could absorb regulatory details day after day, I'd have been a lawyer. A day out at sea could blow some salted air through my brain and help me see things afresh. As something to mull over, I took James's simple statement of the issues then in play. 'The main point is not how you divide up the catch. The main point is how many fish do you take out of the sea; how many fish do you leave in the sea? Only if you get that right does anything else matter at all.'

The pre-reform fisheries policy allowed for a scientific committee to guide the Commission on how to set the allowable catch. 'But then the Fisheries Ministers would get together,' James said, 'and they would set an actual limit that could be 40 per cent more than the scientists said was the maximum that could be sustained. We were trying to make sure that the fundamental principle of the new law was that the maximum sustainable yield of fish was determined by good science, and that that scientific assessment should determine the quotas.'

A road bridge connects England's southwest coast to Hayling Island, flat and ancient land that rears itself above the sweep and tug of an estuary. I cross the bridge for a day out at sea with Harvey Jones. Once a businessman with a dotcom fortune, Harvey swapped that business world for a life of engaged philanthropy, and a boat or two. The larger of his boats is moored in Scotland. The smaller one is still ten metres, a handsome size in these coastal waters. We boarded *The Chalk Hill Blue*, chugged away from the Hayling Island marina, pulled up the sails and cut the engine. Public interest law groups work for free, but someone of course has to pay their considerable costs. Harvey is a committed supporter of marine groups including ClientEarth. I looked for insights of what drove such an investment.

Harvey and his partner dealt in bicycle accessories. They scaled up to go online as Wiggle.com in 1999. Ten years later, he stepped back from its day to day running after a buyout. He and his partner salted away money in a foundation to take care of the seas, and Harvey took off in that larger boat for a break in Arctic waters. In the run of English waterways that leads into Chichester Harbour, he is back in the landscape of his childhood.

'In the summer, all the mud was full of cockles,' he recalled. 'When I was a kid, I had a cockle round instead of a paper round. It was a really bullshit stupid way of catching them. In the sand, there's a little dip, but in the mud you can't see the dips, so you're raking them up with your fingers and you get cut to ribbons on the broken shells. Put them in a basket, wash them. Me and my brother cycled them home on an old butcher's bike. Cook them up in baby Burco boilers, sieve them through so as to get the flesh, put them in half gallon jars in vinegar, and sell them to the pubs so they could sell them to anyone who was drunk enough to want to eat them.'

He laughed as he told the story, and a grin's never far from his face on these waters. He's a lean guy with an old salt's beard, pleased to be out of the mayhem and free on a boat before he'd even touched 50. The freedom had not rendered him romantic. A fan and supporter of Greenpeace, he pulled back from funding their artisan fishing project, which sought to ban big ships and bring in little boats in their place.

'Going smaller doesn't do it. My family locally extinguished species of fish with a boat you could fold up and put in the back of a van. It's nothing to do with the size of a boat. Icelandic fleets have huge trawlers out there, each of them with a government scientist on board, and if that scientist sees a lot of young cod he'll close that area. From a cod point of view, I'd much rather have that than a thousand lunatics out there in rubber dinghies catching every last cod.'

Winds had the boat scudding through the water as sun shone. We barrelled past Thorney Island, with its church that has kept just above water level since Saxon times. The concrete ramp by the sailing club used to be a fisherman's path, where the Jones family harvested their oysters. Their flat bottomed boat sat flat on the mud. They would push the stern one way and then pull it the other to break the suction, and set the boat rushing into the water with a tremendous splash.

Harvey's breakout trip taught him that the fishing methods applied by his own family in childhood, his own tribe, had a universal resonance. 'On a trip to the Arctic, it struck me that most of the stuff we are being told in conservation terms isn't true: "Traditional societies look after their world, we need to go back to tribal culture." It's complete bollocks. People wipe stuff out — full stop.'

Harvey had an early life among those who scraped a living from the sea. That experience informs his current actions. He put it simply. 'You can't get people to make changes for a positive future if all the intermediate stages see them worse off than they are now. No one's going to make themselves worse off.'

That's the fault line that he sees running through all the fisheries projects mooted by NGOs. They will all make fisheries worse off for a period of time. 'And they wonder why people are laughing at them. You need to make sure next week's income is better than this week's income. These are poor people. You need these people onside, or they would undo it anyway. It's a virtuous circle. You might as well be nice to the people at sea and get them onside, because if you don't you're going to fail anyway, and it's a good thing to do as well.'

That's the same reason he and his trust chose to fund lawyers at ClientEarth. 'We use the law not just because it's a good tactic, but we believe that society governed by law is what gives one person one vote, not one dollar one vote. That's what we want to see: a society governed by law. We use law because we think it's a powerful tactic and it's aligned with our aim.'

Harvey brings his childhood to the table, and he brings his subsequent successful business career too. 'It was quite funny to start with,' he recalled. 'I was talking to Greenpeace and they'd say, "We all want to understand these business things." What they really meant was give us the cheque. I said, "Well, OK. What I learned in business is you need a plan, you need some tactics, and the tactics need to line up with the plan, and you need to alter things as you go along if they don't work, and you need to know whether you're succeeding or not. It's not rocket science, but that is what businesses do." "Oh, well, things don't really work like that in the not for profit sector," they said. Too bloody right they don't!'

He remains fond of Greenpeace. 'They're good at cavalry, but if you're fighting a war you don't put the cavalry in charge of strategy on Day One. Each superhero has their superhero skill, but you need a bunch of them.'

What are some other superhero skills?

'ClientEarth are helping build the rule of law in a country. Once you've got it, you can start using it. If you haven't, then nothing's going to work. My family broke every rule going. Voluntary rules meant you completely ignored it, and actual rules meant you just had to be a bit more careful. That's how people are. It's certainly how fishermen are and how anybody doing sport in the natural world is. You're a hunter. It's not bad; it's just what you've got to work with.'

His solution to the fisheries issue is drawn from his business experience. 'It's all management. You can solve any of this stuff if you've got a sensible game you're playing with the resource. The problem is technology more than people. People have always caught fish, but if they didn't have the technology to catch them efficiently enough, then they would always leave some. As soon as the man's got the gun, the wildlife's wiped out. In fisheries, it's the engine. As soon as you've got an engine, it's just too easy.'

His boat was doing without an engine, with Harvey snacking his way through a tubeful of Pringles with one hand and adjusting the wheel with the other as wind filled the sails. We'll let him get lost in his sailing for a while, with just one parting sally.

'Marine conservation is fisheries management. Everything else is commentary.'

The Union Jack that flies on Hayling Island also flies thousands of kilometres away over the Falkland Islands. The Falkland Islands Government is the regulatory body for its fishing.

Fresh with her Masters in Marine Science, Policy, and Law from the University of Southampton, Melissa Pritchard crested the dark Atlantic waves around the island. She was now a scientific observer aboard civilian fishing vessels. The intended catch was squid and blue whiting. About 36 tonnes of squid were scooped up in the nets, but the same haul (because the fish were inseparable) included 34 tonnes of rockfish. These rockfish were small and did not produce good fillets. Sooner than land them, the fishers threw them overboard.

Albatross swarmed after the glut of food. Melissa held on to the railings of the boat and watched. Recent measures had reduced the albatross bycatch by 97 per cent, so they were no longer dying faster than they could breed. No way, though, could they gulp down the tonnage of fish dumped around them. Why did these rockfish have to be thrown back into the water? Couldn't they be eaten?

Melissa's degree courses had imbued her with a sense of the derelict nature of the Common Fisheries Policy. The CFP seemed like a clapped out car; 30 years old and patched together every reform period to keep it going, it was hardly roadworthy. She now had firsthand experience of the senseless discard of fish to add to that distrust of the CFP in its current mode. With the new CFP reform five years away, Melissa joined the three person fisheries team at the UK's Marine Conservation Society.

As part of that team, she was pulled into a planning session for the celebrity chef Hugh Fearnley-Whittingstall. His hit TV show *Chicken Run* and the subsequent 'Chicken Out!' campaign fought to raise consciousness of chicken welfare among poultry retailers and consumers.

He needed his next campaign.

His media team had picked out three issues. For Melissa, one of the three was a standout contender: discards of fish. Iceland and Norway, independent of the EU, had effectively banned discards. All the fish those countries caught had to be landed and recorded, which from a scientific standpoint is crucial. As step one in the environmental lawyers' action plan, such capture of data is vital for any true legal advance in the control of fisheries.

'Nobody likes discards: fishermen, politicians, consumers,' Melissa recalled. 'And it was possible to end them, especially with the CFP reform coming up. So that was the start of "Fish Fight". Some other NGOs agreed. They saw it as even more of a win than "Chicken Out!" Some farmers will farm non–free range chicken, but no fisherman wants to discard. Everybody wants to get rid. How they do it is the contentious bit. Journalists knew what would work with the public. They were really good at translating how awful this problem is, and what you can do about it. I helped them all the way through.'

Through eight hours of television, viewers were encouraged to head online and sign a petition. Melissa came up with the idea of a boat that displayed a digital counter. Moored on land outside the European Commission, whenever someone added their name to a petition to end the discard of fish, the number on the counter clicked higher. Soon there would be over 870,000 signatories from 195 countries.

'It comes down to politics in the end,' is Melissa's verdict. 'The reasons fish stocks were overexploited in the past is political, and nothing to do with science. There was never anybody influential enough to point out to fisheries managers that drastic change was needed.'

Stopping discards became one goal of the fisheries reform, but the issues were more complex than that.

June 2012, and Fisheries Ministers were due at a European Council meeting in Luxembourg. They would be considering the fisheries reform as proposed by the Commission. Some clauses might play badly with the electorates in ministers' home territories. Could they not simply mull over the contentious issues for another decade? Water down the resolution, strip it of binding language, and let that be the version they signed.

Harvey Jones was on board *Saxon Blue*, the larger of his boats, cruising Scottish waters. 'I got a call from the guys from "Fish Fight",' he recalled. They wanted to place full page advertisements in the Spanish, Polish, French, and British newspapers, filled with a headshot of the country's Fisheries Minister. These countries were the major players in the debate. The newspapers would come out on the morning of the Luxembourg Council meeting. Would Harvey pay for such a campaign? 'I said I'd take part if ClientEarth thought it was a good thing to do. The real feeling was that the ministers would sneakily have this meeting they hadn't announced and undo all the good work; there was an underhand thing going on, an anti-democratic stitch up. They were all going to get together and quickly agree everything without anyone having a chance. The adverts were an insurance policy. We wanted to do something to make them think everyone was a bit more on the ball than they might have hoped.'

Sandy Luk was on vacation. You think you'd be safe with your family down a salt mine, but this was Germany. The phone signal got through. She sat in the sunlight outside the cave's entrance, and already her mind was beginning to write the copy for the adverts.

The slogan was to read 'We are watching you!' Who was that 'we'? All those hundreds of thousands who had come on to the 'Fish Fight'/ ClientEarth website to sign the petition against discards. Sandy and her team's text laid out the public's minimum requirements from the ministers' meeting. The whole campaign took 48 hours to come together, from first floating the idea to having it set for the newspapers' editions.

Fisheries Ministers were not used to headline coverage. The newspapers sat beside their breakfasts in Luxembourg that morning. Word of the reaction leaked back to Harvey. 'The Spanish minister was absolutely apoplectic to see his face in the paper. I thought that was quite nice, to just turn the tables on them. The demand was quite modest. It wasn't dramatic or outrageous; it was just, "Do your job. That's all we expect." I thought that was really English and underplayed.'

Fisheries Ministers found themselves locked in 20 hours of intense negotiations, into the small hours of the following morning. In the face of stiff opposition from France, they came out with an agreement to end the practice of discards. They also set a goal of sustainably managing fish stocks, with a deadline of between 2015 and 2020. The path remained open for an ambitious reform package to be achieved in Parliament.[24]

As rapporteur on the policy reform for the European Parliament, Ulrike Rodust worked to rally members from the different political groups as well as from different member states behind what ClientEarth's team pushed as the goal of the reform: setting the catch at the maximum sustainable yield, and making it legally binding. Essentially, this means that fish are not pulled from the seas faster than they can replenish themselves by breeding. Rodust endured years of backdoor quota deals between the member states' Fisheries Ministers. Now 'it was clear to me that senseless overfishing had to be stopped and that the European Parliament with its new powers from the Lisbon Treaty could play an instrumental role in this. NGOs did an amazing job in convincing MEPs inside and outside the fisheries committee of the need for a new policy.' Even so, it seemed that the fisheries committee was set to vote against the new draft proposal.

'A lot of the bad decisions about the environment are not made out of evil intentions,' James reflected. 'They're made out of ignorance, which can then be manipulated by entrenched economic interests. Maybe the people who were about to make the wrong decision on setting maximum sustainable yield were just ignorant? We decided that the last best hope was seeing whether this was one of those rare moments when reason could prevail in the political process.'

The lawyers joined with the rapporteur to arrange a meeting in the European Parliament for members of the fisheries committee.

ClientEarth flew in five top fisheries scientists and fisheries managers for the day. In Parliament that evening, Stanley Johnson, the father of the UK Conservative politician Boris, was the affable Chair at the head of a huge oval table. Scientists lined up together, and MEPs were dotted around. James began the evening by delivering the overall legal perspective, and then passed the argument on to the scientists.

An aide to an MEP approached the ClientEarth team afterwards. Her MEP represented one of the largest and most crucial of those states with active fisheries. 'This event was really important,' the aide reported. 'He's changed his view. The MEP said to me, "I had been told by the industry that it's all very complicated and that politicians shouldn't interfere, it's all working just fine. But what top scientists in the world on fisheries are telling me tonight is it's very, very simple. You leave more fish in the sea and there'll be more fish in the sea. So my goal is to leave more fish in the sea."'

'It was a big experiment from my point of view,' James recalled with some satisfaction. The gamble had paid off. 'Can reason prevail in politics? The answer is yes, when it's done in the right way and at the right moment.'

According to Rodust, the session 'underlined that, among scientists, maximum sustainable yield is a widely accepted "mainstream" concept. This surely helped to convince some MEPs.' Things stayed tight, but they had swung just enough. 'The majority for a reform was very narrow in the fisheries committee (13 for, ten against). Only when the whole European Parliament was allowed to vote on the reform did a clearer majority emerge, thanks to the enormous lobbying effort of NGOs and citizens from all member states.'

Chris Leftwich has been in charge of London's fish market, Billingsgate, for the last few decades. His white fishmonger's cap and apron were firmly in place by the opening hour of 4am, and by 6am he was set to lead myself and a group of young fisheries professionals on a guided tour of the market. So long as we didn't set him off on the subject of discards. There was so much to say, we would probably see no fish.

Well, all right then, just a snippet. 'There's a story about boats going out on the south coast,' Chris began. 'They landed a massive amount of cod because the cod had come in in a shoal. They hadn't got a quota for it, so they phoned the shore and said could we land and give it away to charity? "No," came the reply, "you've got to dump it. If you land it, we're going to prosecute you."' Chris's eyebrows rose and his voice with them as he voiced the fishermen's response. '"We don't want any money for the fish. It's ridiculous. The fish is in the nets, it's on the deck, it's dead. Surely you could give it away to old people's homes." No.'

And then came the punchline.

'You've got to believe that the same happens everywhere throughout Europe, but we know it doesn't. The levels of controls are different in different areas.'

Chris led us through a five hectare complex, where 98 stands and 30 shops shift around 23,000 tonnes of fish a year. Forty per cent of that is imported, though the UK imports about 80 per cent of its fish on the whole — surprising figures for an island nation. Europe simply cannot supply itself with all the fish it needs.

The first fish Chris picked up to show off was a hake, steel grey and glossy. 'The minimum landing size for hake is 30 centimetres, but only if it's checked. With global warming, just a half to one degree change in temperature causes fish to migrate. The hake is normally a southern fish; we wouldn't have caught it past the Channel. Now they're catching masses in Scotland, but they've got no quota for it. So hake is now considered what we call a choke species in Scotland. They're catching it, they use up their quota, and then it stops them catching other fish.'

On a January day in 2011, the UK Fisheries Minister Richard Benyon was pulled into Billingsgate for a 5am TV shoot. The TV chef Hugh Fearnley-Whittingstall was primed to grill him. 'He was a total nightmare to deal with' was Benyon's abiding impression of the chef that morning, 'because he has an ego the size of the Albert Hall and he wanted a minister in a suit to be the ass in every broadcast that he did.'

Benyon's constituency is in West Berkshire, about as far from the sea as England can manage. He fought two losing elections to become its Conservative MP before winning the third. Once a soldier, he learned farming at the Royal Agricultural College in Cirencester. That came in handy, since as a child he was moved into Englefield House, a stately home that has provided location shooting for such films as *X-Men: First Class* and *The King's Speech*. Inherited from a relative who died without direct heirs, it included nearly 6,000 prime agricultural hectares. Richard Benyon is reckoned to be Britain's wealthiest MP; his own declared sense of land ownership is one of stewardship.[25]

In their first interview, the chef challenged the minister on discards. 'He seemed to be expecting me to somehow defend it. I said to him, "I'm a farmer. What would happen if half the calves, cattle, sheep that we slaughter were dumped on the side of the M1? There'd be riots in the street and I'd probably be rioting, demonstrating with them."'

Of course, you don't get to cheat a TV presenter of a moment of drama. The chef laid out 12 fish and set the minister the challenge of identifying them. River fish, well maybe he would have stood a chance. Sea fish? He named three.

'I find him the most frustrating man I've ever dealt with. But he did me a favour,' Benyon accepted, reflecting on Fearnley-Whittingstall's work against the practice of discards. 'You know, he popularised it, and he took the battle to Europe, and all credit to him; the "Fish Fight" campaign was an enormous success.'

I met with Richard Benyon in the UK's own new parliamentary building Portcullis House. He came out to rescue me from the lobby and lead me through to the café tables in the atrium. A general election was due. He was engaged in filtering a green agenda into the Conservative Party's manifesto. I was glad he had found time for me. Governance of the seas evolves through many layers of power and influence, with ministers engaged in the highest tier. I was tracing the effects of environmental lawyers on the whole process. Were they subtle, and therefore invisible, or were they noticed at the top? My opening question involved ClientEarth. Did he know them?

'A highly respected organisation,' he responded. 'They were sometimes "agin" me because that's what they do. That's fine. But they work well with other NGOs. I feel firmly rooted in what they are seeking to achieve, and feel that they have a very important place in the way democratic institutions try to drive environmental policy. To have NGOs backed up with more than just emotion and science, but to have a respected legal basis for what they're trying to do, is very important.'

He recalled his own early days as the UK minister out to steer EU fisheries reform. He found a friend in Maria Damanaki, the new Commissioner for Fisheries, but otherwise initially felt himself to be in a minority of one in the Council of Ministers. 'And then I realised I wasn't. There were some quite green ministers, mainly from the Scandinavian and Northern European countries. Germany was a big ally. I particularly got on with the German minister. Then it was a question of finding friends and cooperation amongst those countries you wouldn't normally see as the usual suspects: France, Spain, Italy, the Mediterranean countries, and the landlocked countries.'

A point of revelation came in his first negotiation one December, when an official came to talk to him about mesh sizes and other technical measures for the Northwest Scottish demersal fleet.[26] He thought to himself, 'This is Kafka. I mean, this is absurd. This man knows less than I do about what it's like to be on a trawler in a big sea off St Kilda.'

That drove the minister to think in terms of a three pronged axis. 'One, devolving power so you don't have officials in Brussels micromanaging the technical measures that the fisherman will be using a thousand miles away. Absurd to have one system that tries to micromanage fishing from the sub-Arctic in the north North Sea down to the Mediterranean waters south of Malta. Trying to break that behemoth up was a priority.

'Two, the discard ban because that was the lightning conductor.

'And then three, which I think is the big win; history will write this as the moment that the whole system turned, which is the legal requirement to fish to the maximum sustainable yield and making that a legal requirement. That's the priority for me.'

Of course, all of that almost slipped away when the Council of Ministers met for their meeting in Luxembourg. How was it to find his face staring out at him from the 'Fish Fight'/ClientEarth full page adverts?

'It caused my family and friends mirth. My late father practically choked on his breakfast when he was reading his *Telegraph*.'

He didn't feel the need for the prod to good action himself, though was used to the presumption that he was anti-environmental. 'There was a slightly bizarre incident in Luxembourg,' he remembered, 'when I was prevented from going into the building to vote for good, green reforms on sustainability by a group of Greenpeace people linking arms around the building. I said to them, "Oi! You want me in there," and I didn't think that they really knew what they were there for. But others did and were a very effective lobby.'

The adverts spurred Benyon to a further reflection. 'When you're in your ministry or going out talking to people, you're aware of your domestic circumstances. The moment you go under the Channel and you enter those vast edifices, the Commission, the Justus Lipsius building, where most of our negotiations happen, it's so easy to feel detached. What you do hear from is quite a well organised industry representation. So it is good to be reminded that this is an important issue for many more people than the industry. And it allowed me a bit of capital to say to the industry, "Look, this is a big popular movement. If you don't agree with what they're saying, you've got to manoeuvre and articulate and debate with them. You can't just come into my office and whinge. I happen to think these people have got a case and I want to back them up."'

Law, as we have learned, is ultimately a thing of language. Ulrike Rodust, the rapporteur for the fisheries reform, recalled how 'all major NGOs provided proposals for amendments to the reform, which were useful when drafting my report for European Parliament'.

As legal scholar William Robinson notes on EU legislation, 'The Commission's proposal is the work of many hands and minds.'[27] Among all those interested parties, 'the first drafts are generally produced by technical experts in the department responsible. They are generally not specialists in legislation ... Only very few DGs [Directors General] have, within their own legal units, lawyers to help their technical experts with legislative drafting.' Furthermore, almost all first drafts 'now have to be produced in English. It is a real challenge for technical experts to have to draft complex texts in a language which is not their mother tongue.' These drafts then go to the Commission's Legal Service for vetting, by a staff of almost 400 lawyers, who are divided into teams which specialise in different fields of EU law.

Imagine their relief when language appears before them which is charged with legal competency. As the CFP reform process staggered towards its deadline, ClientEarth lawyers provided input to group after group. Robinson notes the

> large numbers of interveners, all pursuing their own agendas. Those agendas may differ considerably as between the 29 Member States and the 750 Members of the European Parliament, most of whom are attached to one of 7 political groups (but 52 of whom are 'non-attached'). Only a minority of those interveners are specialists in legislative matters and yet they have to interpret the legislative text in the proposal and suggest textual amendments to it.

And when doing so they are mostly working in a foreign language.

Melissa Pritchard, from her own English language, science background, watched the proceedings in awe. She switched organisations so as to be the scientist on the ClientEarth fisheries team, and Melissa was speaking retrospectively of that role when we met. She saw the play of law from a close angle.

'I'm really amazed at how much is in the final law; those exact words came from our mouths and we wrote them down,' she recalled. 'If we'd just been a science organisation, that would never have happened, because we don't know how to write the law.'

She thought back to how it had come about.

'Once you've got a draft of the CFP, every clause can be worked on, so we worked with lots of partners. These politicians don't understand the science and the law, that's not their job. As I learned at ClientEarth, you can read one thing, but lawyers have to point out what it actually means. Things like "may" is different to "shall". Maximum sustainable yield, deadlines, limits, understanding what all that means — we had to educate the politicians and conservation groups, especially on the legal aspects. We worked with so many groups to put the amendments through. That's why the majority of the wording is wording the groups will recognise as what we all fashioned together. It was a huge advantage having the lawyers and scientists working together on that.'

Back in 2006, we saw Professor Boris Worm's gloomy views of fisheries collapse by 2048. In 2015, he co-wrote of how some optimism set in. 'Countries formerly at war began to work together to hammer out new deals for fish, exemplified by both the recent revision of the Common Fisheries Policy in Europe and new efforts underway at the United Nations to better regulate fishing on the high seas, the 60 percent of the oceans outside national control.'[28]

A 2016 report published in *Proceedings of the National Academy of Sciences* somewhat challenged such optimism.[29] 'It comes from a group of American scientists who have been quite outspokenly optimistic about the future of fisheries and this study confirms that optimism may be OK in some regions like our own but not so much elsewhere,' Boris Worm said, having been asked to add an official commentary to the study's print version. 'If you fish these stocks the exact same way you're fishing them now and you keep that up, then indeed we will face in 30 years or so a world where according to this study almost 90 percent of stocks are depleted.'[30]

The reform of the Common Fisheries Policy was a fair victory for those with a long term interest in keeping fish in our seas, and it was won against the odds. Gains made, as ever, are open to reverse, and vigilance has to be forever.

It reminded me of James's dictum: Winning is good but temporary. Losing is permanent. You have to keep fighting.

Harvey Jones skippered his boat in his estuary, but his eyes were set on a distant horizon: the next fisheries reform process. The recent round was more successful than he had hoped, particularly in its introduction of maximum sustainable yield. 'Clearly chucking usable fish back in the sea is bonkers,' he allowed, then went on to enumerate some of things he would like the lawyers to sort out next time around, 'but the issue is how many you killed in the first place, not what you did with them afterwards. You need fully documented fisheries; you need to be managing by results not technical measures; you don't measure my mesh size, you measure by how many fish you kill. Your scientist doesn't need to be sitting on a boat, but he needs eyes and ears on a boat, and technology can provide that. It's an excess mortality ban you need, not a discard ban.'

As his own hands on example of effecting change, he is running a scallop ranch further east, in Lyme Bay. That is in a special area of conservation, as is the estuary in which he was sailing. These marine areas were protected because 'ClientEarth forced the UK government to implement laws they had signed on to — really clear, well written EU laws on what you had to do,' Harvey reflected. 'Other people I've spoken to have said this is the most important thing for marine conservation that's happened in my lifetime.'

'An official came into my office and said, "Erm, I think you might need to clear some time in your diary, Minister; something's come up,"' Richard Benyon recalled. 'Which usually means something pretty serious. ClientEarth stood on the head of a legal pin and got a favourable adjudication.'

That 'adjudication' did not come from a judge or a court, but from independent counsel brought in by the minister. Counsel confirmed everything in ClientEarth's letter: that the government was in breach of its obligations under the Habitats Directive, the EU's flagship of environmental regulation, that the UK transposed into its own law in 1994.

At the time, the NGO community was examining how to put together a new environmental planning law. Such moves take years, they work or they fail, and in the meantime destruction continues. ClientEarth teamed up with the Marine Conservation Society to fuse together their joint legal and scientific expertise. They looked at sites designated for special European protection that were routinely fished and even dredged. And they pointed out how this was in total contravention of two articles in the Habitats Directive.

'They should have stopped all fishing activities and then permitted the ones sequentially when they passed these tests,' recalled Jean-Luc Solandt, the Marine Conservation Society's lead on the case. 'We knew that would destroy the industry in large parts of the country, and just create too much bad will, so we agreed to a phased approach.'

A trip down to the southwest coast of England helped me understand how this application of a neglected law brought immediate environmental benefits.

Brixham invented trawlers. The red sailed fleet dominated the English Channel, and Brixham sailors took the trawling concept to the eastern ports of Hull and Grimsby. The town's fish market still comes top in Britain in terms of value. The nation's top buyers come for the 6am auctions. They gather around the auctioneer as he moves between the white tubs of fish. Brief nods, shakes, a twist of a wrist, a muttered word, and the catch is sold, its price relayed to a digital display in each hall.

Tim Robbins shared nods and words with many of the men. He was not there to buy but to inspect, in his role as manager of the local Inshore Fisheries and Conservation Association — IFCA.

Tim had left school at 16 when he spotted an apprenticeship on the local patrol boats. He'd spent 35 years working this bay since then. His job changed after ClientEarth and the Marine Conservation Society's legal challenge. The years since 'have been really different. I can be proactive rather than reactive. It's down to what ClientEarth and the Marine Conservation Society did. Defra [the UK governmental agency] wouldn't have budged. We'd still be pottering around the edges.'

He manages the system of phased protection. A vast chart in his office is dense with rectangles coloured red, amber, and blue. Each rectangle is a zone, and the colour designates the urgency of its protection. Tim, Jean-Luc, and I boarded a rigid inflatable that acted as that afternoon's fishing patrol boat. We motored out into Lyme Bay in the red zone of the matrix, an area that already received protection.

'We have coral reefs,' Tim explained of the bed below him, 'but it's a cold water coral, so it's very slow growing. Obviously, if it gets hit by a couple of tons of steel that's part of a fishing instrument and gets pulverised, it's going to take a long time to recover.'

Jean-Luc leaned on the side and stared down. 'Seagrass,' he explained. Beds of the stuff grow along the bay up to the five metre contour. They're vulnerable to damage from towed gear. 'So what?' I wondered out loud. Coral I understand, it grows so slowly, but surely seagrass can restore itself.

Jean-Luc stared, dumbfounded for a while. He expected better of me. For one thing, the grass was a nursery for fish, he explained. 'It's also really good at stabilising sea bed habitats from winter storms, and it's good for carbon capture and nutrient recycling.' If I needed more, he could give me more. He dives these waters. The more you come to know, the more you have to learn.

Tim spent years working to convince Defra to support a switch away from bye-laws to a flexible permitting system. They would not listen. Suddenly, after ClientEarth's intervention, his plans got the green light. 'I think the challenge put the pressure onto Defra to consider properly what we were trying to do, so it accelerated the process,' Tim told me. 'They might have kept prevaricating around it, making us dance around a bit more.'

Mussel lines held up by buoys mark the border of the European Marine Site. These are the current home to Harvey's juvenile scallops. Tim's words remind me of Harvey as we pass by. Fishermen are poor, Harvey said. You need them onside, or they'll undo all you're trying to do.

'They don't like any restrictions and won't engage with any bureaucratic process,' is Tim's take on the fishermen, and he smiles. 'I only hear from them when changes set in. They're grumblers. They all settle down after a while. Eventually, they see the benefits. I get used to the ebb and flow of grumbling.'

Bye-laws meant imposing blanket closures. If the fishing industry didn't like it, then they would disobey en masse. Tim didn't have enough policing power to protect the sites. The flexible approach means he gets their buy in.

'Fishermen are hunter gatherers,' Jean-Luc agreed, 'free in an environment where they want to be free. When we get into rooms together, I often like fishermen more than NGOs. My grandfather was a harbourmaster and a sailor, so I don't feel uncomfortable in the company of fishermen. You can actually have a laugh and you can see commonalities.'

Jean-Luc is a scientist out to preserve the marine environment. His work with ClientEarth added a whole new legal approach. 'We're taken seriously when we have an opinion' was how he summed up the difference.

The boat revved to 27 knots, kicked up white wake, and bounced across the bay. Jean-Luc grinned. Yes, he loves the sea, and now above the roar of the boat he had to shout to be heard. 'I'll defend it to the death using law.'

Setting up in Brussels

James Thornton

From the beginning, I knew that ClientEarth needed to be in Brussels. The EU countries go to Brussels to make the laws. If you want to influence them, you have to be there.

The first step is to get the text of the law right. Translating policy into law is within the lawyer's art. This should perhaps be obvious. But lawyers were not helping the NGOs write laws. There was not one practising lawyer in any of the environmental NGOs in Brussels.

One of my early meetings in Brussels was with a Member of the European Parliament who was then Chair of the Environment Committee. When I explained what I wanted to do, he welcomed it. He said that when an environmental bill came up, he was lobbied by industry, who were represented by sophisticated lawyers. They arrived with a package of amendments for the bill and arguments about why they were good for the economy. The environmentalists, he said, do not have this kind of representation, and so the contest is very uneven. 'And I,' he added to my surprise, 'also have no legal help. There are no lawyers on my staff, and I am not a lawyer. We need your help.'

There was clearly a need. How could I fund the office? Kristian Parker of the Oak Foundation played a key role. He was one of the earliest funders of ClientEarth. He saw the need for a Brussels office and would fund one position for two years. This should let us demonstrate our value and secure other funding.

There was a hurdle to get over before I got the funding though. The Oak Foundation put my proposal out to a peer reviewer. The response they got was that they should not consider funding ClientEarth. The reason? I was proposing something unsuitable. Americans are hyper-aggressive and sue everyone. Europeans on the other hand are much more polite. We talk our way through problems, the reasoning went, we don't need this aggressive newcomer. Everything is fine as it is. Don't encourage him.

I heard the same argument often in the early years of building ClientEarth.

The Oak Foundation gave me the chance to respond. I spent an intense week writing a 30 page paper arguing that legal expertise including litigation was needed in Europe. Scholars were arguing for it, because there was an enforcement gap, where good laws were passed and then forgotten. Enlightened activists and parliamentarians wanted the help we could provide. Moreover, as I pointed out, it was only civil society that was lulled into not defending their rights. Companies in Europe were just as aggressive about using the law as companies in the USA. Why keep citizens unprotected out of a naive sense of politesse? Ludwig Krämer, the leading expert on EU environmental law, added a supporting statement.

The Oak Foundation accepted my argument. I put out ads looking for a lawyer. I met a young French barrister called Anaïs Berthier. She worked in the Brussels office of the top French corporate environmental specialist firm. She was clearly talented and also dedicated. I was honest with her that I was looking for someone with more experience. 'That's all right,' she said, 'I am not going away. I will wait until you can hire me.' This made an impression.

Some months later, Anaïs became my first employee in Brussels. So she wouldn't be on her own, we rented a desk in the WWF Belgium office. She started in on access to information and access to justice issues.

Transparency and accountability are vital to democracy. If citizens do not have access to information, they cannot participate in decision making. If there is no way to challenge government actors when they violate the law, there is little incentive for them to comply with it.

I am a great fan of the European project, and ClientEarth is fully supportive of it. The standard of environmental laws is much higher across all European countries than it would be without the EU. Environmental protection is one of the best things that the EU does. But unfortunately, the European institutions often do not make themselves easy to love.

When it comes to transparency, the EU institutions like to say they are the most transparent in the world. There is a law requiring the release of documents to citizens upon request, with limited exceptions. When citizens request such documents, however, they are routinely denied. You may appeal the denial in an administrative process. This is difficult for citizens and NGOs to do, and so we often help them. Even when an appeal is made, the EU institutions routinely deny access to the key information that would let the citizen challenge their behaviour.

The one route to the European courts for citizens is to challenge the denial of your request for information. The European courts have granted citizens access to documents, though their recent trend, moving in the wrong direction, is to restrict access. Nevertheless, we are still winning cases, and continue to push the EU institutions to comply with their own law, and to live up to their rhetoric on transparency.[1]

What about access to justice? How easy is it for EU citizens to challenge the EU institutions when they violate their legal duties? The answer should be that it is easy and even welcome. Unfortunately, it is not the case with the EU. Though the foundational treaties of the EU grant the right of access to justice, the European courts have barred the doors to citizens, denying them standing.

In many legal systems, to have standing to sue, you need to show that you have an interest in the outcome. This is expressed in different ways, such as having an individual interest, or suffering an injury.

The courts of the European Union have worked hard to deny citizens standing. They have dipped into logic that would make sense only to medieval theologians. The test for standing in the treaty is that you must have a direct and individual concern. This should be read broadly so as to allow citizens to test actions of the EU for legality.

The way the EU courts interpret 'individual concern' is that it must be a unique concern. Consider what this means. If the EU is violating a law, say authorising more fishing than the law allows, it affects all EU citizens. Fish stocks will suffer. The harm affects more than one citizen. Pesticide laws, air pollution laws, biodiversity laws, these all affect more than one citizen.

No one has a unique interest in protection of the environment. The interest is always shared. So under the courts' medieval logic, there can never be a citizen who has a right to come to court. No one will ever have a unique interest. The worse and more widespread the harm, the weaker your putative right becomes. It is a perfect way to deny citizens environmental justice.

One of our first actions at ClientEarth was to challenge this restriction on access to justice in the European courts. We raised the issue in the Aarhus Convention Compliance Commitee in 2008 (at the same time that we complained of the cost rules in the UK and standing in Germany).

Anaïs Berthier argued the case before the Committee in Geneva. Our challenge posed a constitutional problem for the EU. They had signed on to the Aarhus Convention, promising access to justice to EU citizens, but were not delivering it. And now they were before a body that had the power to judge the EU itself. This is a unique position for the EU, which is careful to insulate itself from anyone being in a position to correct it. If we won, would the EU allow itself to be corrected by the Committee?

The good news is that, eight years later, we finally got a good decision from the Committee.[2] It found that neither EU legislation nor the jurisprudence of the European courts complies with the obligation to give citizens access to justice. The decision should move the EU to increase the accountability of its institutions, thereby empowering civil society and upholding democratic values. The question is whether the EU will follow the decision, or dig in its heels and ignore its obligations. Only time will tell.

Why have the European courts been so afraid of citizens? They let companies bring cases. The number of citizen environmental cases would be a fraction of the cases brought by companies about their economic interests. By denying access to citizens, in violation of the underlying law, the EU courts make it hard to love Europe, and that is a great shame.

So making the EU institutions improve is a key part of our work in Brussels. We intend to hold Europe to its ideals and its laws.

Another Brussels speciality is trade agreements. As I write, the EU and USA are engaged in a multiyear process of entering into a trade treaty. The devil, as always, is in the details. The proposed treaty, the Transatlantic Trade and Investment Partnership, or TTIP, includes a nasty provision that would allow US companies to attack those European environmental laws and decisions which breach special investor rights. This would be accomplished through secret hearings in a new trade tribunal where the 'judges' would be corporate lawyers.

Canada is even further ahead. The EU-Canada Comprehensive Economic and Trade Agreement (CETA) is currently awaiting ratification in Ottawa and in the capitals of Europe. This agreement contains a similar provision which would allow not only Canadian investors to sue, but also American companies that have Canadian subsidiaries. There is, in fact, a whole string of similar trade agreements with countries all over the world lined up in Brussels.

Would companies bring cases in the Star Chamber that these treaties create? There is no doubt. For example, a US company could argue that the EU's toxic chemical regulations are more stringent than their US counterparts, and therefore constitute a 'barrier to trade'. This kind of attack would reduce protections of health and environment down to the lowest common denominator. The result would be that companies bit by bit destroy the body of European laws to protect the environment and health that have been painstakingly crafted and agreed on for the last 40 years. This behind closed doors destruction of democratically created protections by private power against the public interest is one of the signal evils of globalisation. What can be done?

Here is where it is good to have lawyers on your side. Through the work of ClientEarth lawyer Laurens Ankersmit, we have been able to demonstrate that the EU has no power to enter into a treaty with such a provision, because it fundamentally undercuts the jurisdiction of the EU's own courts to decide all matters of EU law.[3]

After we issued our opinion, the German Association of Judges, representing 16,000 judges and public prosecutors in Germany, and the European Association of Judges, composed of 44 national associations of judges, issued an opinion making the same argument. Over 100 law professors have also signed a letter with the same opinion.

The newest turn in the road is a positive one. The offending dispute resolution mechanism in the Canadian treaty is set to be judged by the European Court of Justice, the highest court in Europe. Its journey to the courthouse steps illustrates the power of legal strategy.

The journey starts with the fact that all EU countries must agree to any trade deal. Belgium in turn has six regional and linguistic community parliaments, which must approve any trade deal before Belgium can agree. So the region of Wallonia in Belgium has as much say in any trade treaty as Germany or France. After consulting with ClientEarth lawyers, Wallonia used this power. Its parliament passed a resolution objecting to the Canadian trade treaty, with their number one issue being the illegality of the dispute resolution mechanism.

Wallonia then used its veto power to bargain. It gave up its veto on condition that Belgium ask the European Court of Justice to rule on the legality of the mechanism. Belgium has standing to do so, and agreed. This review by the Court is what ClientEarth has been trying to achieve, and so the result in Wallonia is that the ClientEarth strategy has now reached its goal. 'What we managed to get here is important not just for Wallonians,' Wallonia's Minister-President, Paul Magnette, declared, 'but for all Europeans.'[4] The Court will almost certainly rule the mechanism is illegal, since thousands of judges have already declared it so, and since the mechanism undermines the Court's own jurisdiction. To preserve their own system and their own authority, the judges on the Court must strike the dispute resolution system down.

The Court's ruling will also apply to the parallel provision in TTIP, the US trade treaty with the EU, which will follow the Canadian treaty.

Thus a legal strategy deployed by a single lawyer at ClientEarth may stop the destruction of 40 years' worth of health and environment law built up by the EU.

When something is simply illegal, and is called out as such, it should be possible to beat it. We've done it before.

One of the few virtues of Brexit is that the UK has been a leading voice for TTIP and handing control to companies. France and Germany are less enthusiastic about letting globalisation rip away protections for their citizens. With the UK's voice diminishing in EU decisions, there is a greater chance TTIP, or at least its worst provisions, may be defeated.

Speaking of Brexit, there is no avoiding the fact that voters in the UK have made a decision to leave the EU. While I recognise the faults of the EU, one of which is that it makes itself difficult to love, we work to help move the European institutions in the right direction. We spoke out publicly against Brexit, pointing out the dangers for the UK environment in leaving.

Brexit brings uncertainty, and it will take years of effort to sort it out. The main point for the environment is that, in leaving the EU, we will need to ensure the UK gives its environment the same standard of care it has been obliged to do. Under any leaving scenario, the UK will not be obliged to keep the nature protection laws that now apply across the entire continent. The UK is one relatively small country, and the problem we face is that it will consume more attention to merely ensure the level of protection for nature and human health that were already guaranteed.

7

Coals of Fire

Two hundred metres high, the smokestack's job was to belch fumes into whatever winds whipped in from the sea. In October 2007, a band of Greenpeace activists entered the compound of Kingsnorth, a coal fired power station on the Kent coast. They scaled the smokestack with cans of paint and began to write a message for the British Prime Minister, Gordon Brown. GORDON BIN IT, it would read.

This particular power plant was already doomed. Europe's Large Combustion Plant Directive of 2001 set limits on emissions from power stations. Kingsnorth was too old to conform. It could continue for a maximum of 20,000 hours. The controversy was about what would replace it. Its owners, E-On, had plans for a new coal fired power plant. So did others. If they were built, the UK would have declared dependency on coal for decades to come.

The six activists were hauled down mid-sentence while their paint still dripped. The drama moved to Maidstone Crown Court, where they were accused of causing £30,000 of criminal damage. The activists' defence was 'that they had "lawful excuse" for their actions: they sought to protect property around the world threatened by climate change,' their lawyer Mike Schwartz recalled. 'Their vivid accounts of melting ice caps, expanding oceans and deforestation — and resulting erratic weather, flooding and rising sea levels — were supported by expert evidence.'[1]

Mike Schwarz was using the same principle of law that allows firemen to kick down a door in order to put out a fire: the campaigners were protecting themselves and the world from the carbon pollution caused by the coal plant. The star expert flown in from the USA was James Hansen.

Twenty years earlier, James Hansen testified at a different hearing. As Director of the NASA Goddard Institute for Space Studies, Hansen told the US Congress that humans' release of greenhouse gases had triggered a process of long term planetary warming. He 'noted that global warming enhanced both extremes of the water cycle, meaning stronger droughts and forest fires, on the one hand, but also heavier rains and floods'.[2] That speech of June 23, 1988 can be etched into the calendar of human civilisation as an epochal date: from that moment, humans were no longer simply liable for global warming; they were complicit. Some might act out of denial, but no one could act from ignorance anymore.

Evidence continued to amass. The International Panel on Climate Change (IPCC) issued its first report in 1990, to reveal an overwhelming scientific consensus around similar findings. 'It is hard not to find the prime villain of the piece,' writes Dieter Helm in *The Carbon Crunch*. 'It is the burning of fossil fuels — almost everyone knows this. What is less appreciated is that all fossil fuels are not equally bad and, of these, coal bears the lion's share of responsibility. Coal is worse than oil, and much worse than gas.'[3]

Balding, dressed conservatively in jacket and tie, James Hansen took pains to lay the facts before the jury. The European Directive set its emissions controls on sulphur oxide, nitrogen oxides, and dust, but Hansen's main focus was on coal's vast CO_2 emissions. 'It will be necessary to return atmospheric CO_2 to 350 ppm [parts per million] or lower on a time scale of decades, not centuries, if we hope to avoid destabilization of the ice sheets, minimize species extinctions, and halt and reverse the many regional climate trends …' he told the court. 'There is just barely still time to accomplish that, but it requires an immediate moratorium on new coal-fired power plants that do not capture and sequester CO_2

and a phase out of existing coal plants over the next 20 years.'[4]

The legal argument won. The jury was persuaded to find 'The Kingsnorth Six' not guilty. 'That was a pretty spectacular case within a campaign,' Matt Phillips recalled. He helped shape and fund the Greenpeace campaign from his position within the European Climate Foundation. 'It was about winning through overwhelming argument and campaigning, making it impossible for John Hutton, who was the Secretary of State, to give the permit.'[5]

The scaling of Kingsnorth's smokestack and the resultant trial saw lawyers take an active part in a campaign. The verdict, as the trial lawyer Mike Schwarz admitted, had 'no binding effect in future cases'. As part of the fight against the UK's coal fired power plants, a legal strategy was unfolding elsewhere.

ClientEarth was a fledgling at this time. Tim Malloch joined the few staff as a litigator.

Tim had broken free from private practice to cycle from Ushuaia in Argentina to Prudhoe Bay in Alaska, which is maybe what a litigator needs to do to let off steam. His goal on joining ClientEarth? To stop coal fired power plants. Tim 'worked in an in depth way, developing deep and elaborate legal theories', Karla Hill, ClientEarth's Programme Director, recalled. 'He worked his legal theory to the point where he thought he could beat the government.

'The more traditional NGO view of how you use litigation is that it's a tactical tool as part of a wider campaign in order to achieve some impact — publicity maybe, or slowing something down. ClientEarth brings a broader strategic approach into thinking through the whole dynamic — what the players were doing, what the government was doing and what the legal overlay was. Kingsnorth was the headline, but in fact all of our legal strategy was directed towards the UK government.'

One of the main tools was Europe's Strategic Environmental Assessment Directive. Tim Malloch moved back to the corporate world, but attached his successes from ClientEarth to his new company's website: 'Tim's most notable regulatory achievements include persuading the UK government (DECC) that in order to comply with Directive 2001/42/ EC it would be prudent to conduct a Strategic Environmental Assessment (SEA) in respect of its plan to grant consent to a new generation of "capture ready" coal power stations. On the same day that DECC announced it would conduct an SEA it also announced a ban on the construction of coal power stations that did not capture their carbon emissions.'[6]

'It was a very satisfying moment,' Karla Hill recalled, 'to see that you could make a change.'

On a separate tack, using the California model James knew from his work in Los Angeles, ClientEarth introduced the notion of Emission Performance Standards to the UK. The idea is to set a standard for emissions per kilowatt hour for electricity generation. If you set the standard correctly, coal will fail, and more benign fuels will be favoured.

Tim Malloch's web entry about his time at ClientEarth logs his other success as persuading 'the UK government to introduce an Emission Performance Standard to limit the carbon dioxide emissions from UK fossil fuel power stations'. How much credit can ClientEarth really take for stopping a new generation of coal in the UK?

'It was a whole mosaic of actions,' is how Karla Hill sees it. 'Kingsnorth had the full spectrum — you had expert think tank analysis, industry insiders and investors, legal strategies, more standard environmental and development organisation campaigns, policy and advocacy work, and you had activism. You can see the whole range of approaches. Get them working in the same direction and you have an incredibly effective way to effect social change. Our litigation contributed to blocking the development of new coal and the turnaround of that policy. It could well have worked without the ClientEarth litigation.'

So what happens when other parts of that mosaic of actions are removed, and the legal strategy of a public interest law group is what remains?

After decades on the periphery of a decaying Soviet system, Poland came in from the fringes. Fifty four Polish MEPs entered the European Parliament in 2004, and Poland became the newest and fifth largest EU member state.

To meet EU entry requirements, government ministries teamed with NGOs to ensure the requisite environmental standards were achieved. Accession brought massive funds to improve infrastructure. Cleaner technologies replaced the worst polluting ones. All might have gone swimmingly but for one factor: Poland's near complete reliance on coal. In 2004, 92 per cent of its domestic electricity came from coal fired power plants.[7]

Poland's then Prime Minister Donald Tusk wrote in the *Financial Times*:

> In the EU's eastern States, Poland among them, coal is synonymous with energy security. No nation should be forced to extract minerals but none should be prevented from doing so — as long as it is done in a sustainable way. We need to fight for a cleaner planet but we must have safe access to energy resources and jobs to finance it.[8]

'Energy security' is a specious claim, since Poland is a net importer of coal, largely from Russia. Its own coal has become expensive to extract now the readily accessible seams are exhausted, and is of a higher quality than nations are prepared to pay for. The state owned Kompania Węglowa (KW), Europe's largest coal mine, suffered a loss of more than 1 billion zlotys ($315 million) in 2013,[9] and a further 1.3 billion zlotys in 2014, with losses continuing to accrue in 2015.[10] In April 2016, six state controlled companies agreed to invest $624 million in the mine to save it from declaring bankruptcy by month's end.[11] The industry is unsustainable without vast subsidies. There is no environmentally 'sustainable way' of using coal unless you capture all the carbon. And without a meaningful price on carbon emissions, there is no economically sustainable way to capture the carbon.

There are other reasons for Poland's addiction to coal. Out of five functioning coal companies in Poland, four are wholly or partially state owned and return losses to government coffers. Ministers and vice ministers leave government roles to take up senior positions in the coal and energy companies. Government officials are therefore personally invested in maintaining coal as central to the Polish economy.[12]

The EU's Large Combustion Plant Directive of 2001 meant all old power plants had to meet with the Directive's emissions targets by the end of 2015 or shut down. In 2009, Poland had 14 new coal fired power plants on the drawing boards. Build them, and Poland would have tethered its future to spewing out excessive carbon emissions.

How do you stop that happening? No campaign group fancied the task. No foundation was prepared to fund a Polish office. James chose to risk funds drawn from an EU LIFE+ grant and started ClientEarth Polska, which in May 2016 rebranded itself as *ClientEarth Prawnicy dla Ziemi*, 'Lawyers for the Earth'. From a Warsaw office filled with Polish lawyers, what could the law do?

It took 14 complex strategies to take on 14 new power stations, so I focused on just the one: Północ, pronounced 'puwanotz'. The name translates as 'north', and that's where it is set to be, 400 kilometres of highway north of Warsaw.

Usually, coal fired power plants locate near coalmines rather than near the end users. It consumes energy to transport something so heavy.[13] Dust loss and spillage occur along the route as well, especially during transfers between two modes of transit, such as ship and rail.[14] Północ is far from Poland's coalmines, which are located in Silesia in the Southwest. Its coal supplies would need to come by rail from Russia, or ship via Gdańsk. There's no use of local resources and no gains in terms of energy security.

It makes no sense to me, but it did to Poland's richest man. Jan Kulczyk was worth $4 billion, a fortune largely derived from Poland's privatisation of state owned businesses in the 1990s,[15] and his company was investing €3.1 billion in two gigawatts of capacity at Północ. The Environmental Impact Assessment (EIA) for the greenfield site was approved in 2011, when the company was also granted its building permit. Building would start in 2012, with the plant operational from 2016.

Stop that if you can.

Pomerania is a good place for wind. I looked out from the farmhouse window and watched the blades of a wind turbine beyond the brow of a hill. A small band of locals who wanted to stop nearby Północ in its tracks was gathered around the dining table.

Monica Stefańska was a young mother. She wanted the air and soil to be as rich and clean for her children as when she was growing up. Her mother, Aniela Kamińska, was to hand, piling the cakes onto individual plates and pouring tea and coffee into the best china. Campaigners led by her daughter had a big fight on their hands. She would bake for every one of them if it helped. She wrote poetry for the cause too. One addressed Północ as 'You big monster', an uninvited visitor who should go away and take its noise and stench of death with it.

> You tantalise us with what's in store
> But it is obviously not true.
> This dust will poison us.
> Heavy metals stay deep in the soil
> And deprive us of the bounty
> Of mushrooms from the forest.
> Do not poison our queen,
> the Polish Vistula River.
> Let her run clean from the mountains to the Baltic
> As she has through the ages.

'The investors in the power plant organised a professional campaign,'

Monica told me, 'which included a newspaper called *Nasza Elektrownia — Our Power Plant!*'

This was shameful behaviour apparently. I caught that from Monica's tone. Didn't I get it? she asked. When people read the word 'Our' in the title, they presumed it meant the government. A new state owned enterprise would seem a positive development to them. They might expect some benefit from that. It was shocking the way they hid that the operation was privately funded.

I was intrigued to learn this take on private capital. I still waited on the villagers' strategy for defeating the coal plant, however.

Mariusz Sledź, a retired teacher of nature studies in the local school system, had his own plate of cake at the farmhouse dining table. He was one of a small band of locals who dared to sign their names to a group opposing the power plant. It seemed that all those in power in the region were on the other side. Two open forums were held in separate villages, only one of which was publicly announced. Investors fronted the meetings. Locals were left confused. The investors did not appear to be competent to deal with their environmental concerns. The trouble was that when the locals gave voice to those concerns, the arguments lost sense in the telling. They were left feeling stupid.

Monica had visited the Mayor of the nearby town of Pelplin and won the opportunity for a fresh debate. This would include experts from both sides. The investors turned up for it, and took the stage in their tailored suits. Their arguments were marshalled on their laptops, yet for all the techno-dazzle of their approach, they seemed unprepared. The ecologists on the stage were scruffy in comparison, but kept a proper debate alive.

The debate was hard to keep abreast of, however, for the ground kept shifting. Hardest was keeping track of the EIA on the 200 hectares of the site. Details seemed inconsistent and confounding.

Perhaps a couple of public interest lawyers from a new group in Warsaw, ClientEarth Polska, could help them sort things out?

Marcin Stoczkiewicz was once a high school environmentalist. Marcin's friends were part of a group that blended the environment and peace as its goals, *Wolę Być*, ('I Prefer to Be' in English).[16] Theirs was a nonviolent agenda, with a focus on stopping the polluting industries of the Soviet era and resisting Russian occupation. When the government planned to dam a river near Kraków, to cut down a forest and destroy a protected area, Marcin joined his friends in the streets.

Freedom from Communism in 1989 helped optimism surge in Poland. In the following year, unemployment soared. Among the first to be spared national service, Marcin found a haven in Jagiellonian University. Marcin graduated in Philosophy and in Law. Armed with that degree, he twinned a law lectureship with work as a partner in a commercial law firm that had an environmental focus. Between 2000 and 2004, as part of the process of accession to the European Union, EU directives had to be transposed into Polish legislation and so the government drafted in the support of the legal profession. Marcin worked on a number of environmental laws, including those steering Poland towards its ratification of the Aarhus Convention. Poland bound itself to Aarhus's three pillars: access to information, public participation, and access to justice.

In 2009, global hopes for action on climate change converged on Copenhagen. The Copenhagen Climate Summit left those hopes deflated. It shook Marcin. He saw two options open up: maintain a comfortable life, or take action. Marcin became the senior member of ClientEarth Polska's first three employees, and began a commute from Kraków to Warsaw.

An early task was to try to block the building of another new coal fired plant, this one located in Opole, the capital of the coalmining region of Upper Silesia. His opposition to that development saw him pinned beneath the lights on primetime TV. Poland's Treasury Minister had branded ClientEarth 'an enemy of the State'. Interviewers demanded to know his response.

Later, his personal phone, to which only his family had the number, rang. Instead of the voice of a loved one, he heard the shots of gunfire.

Another time, police charged into the ClientEarth Polska office and forced everybody onto the street. There was a bomb threat, they explained. The sniffer dog that might help in such an instance was left outside.

And then the leading Polish business newspaper branded ClientEarth 'eco terrorists'.

Marcin's position is calmly stated. It shows why the group of citizens around a farmer's dining table, working to form their opposition to a new coal plant, was important to him. 'We live in a democratic country,' he told me. 'Government should take into account different voices and points of view so there is space for civil society organisations. Our role is not to be oppositional to government, but to raise important things for policy thinking and the legislative process. It's not a matter of black and white, good guys and bad guys. There is a lot of pressure for a government to take economic interests into account, but there are different interests and perspectives too. These need to form part of public debate.'

Północ had its building permit. Marcin combed it for anomalies. ClientEarth submitted a 30 page challenge. The developers set one of Warsaw's leading and most expensive law firms on the case. A 300 page letter came back in response. ClientEarth's team responded in kind. Volumes of argument and counter argument emerged.

'The role of lawyers and NGOs should be underlined,' Monica Stefańska told me around that farmer's table, 'because no local community can do anything without professional knowledge. People in the local area don't have money to afford professional lawyers.'

It's true: billionaire developers can outspend a local community when it comes to legal challenges. And environmental law is stacked with technical issues. I have chosen to look at efforts to stop just one of the 14 proposed power plants in Poland. Even this one battle to halt Północ is being fought on several frontlines, and I will examine just one of those. Take time to consider, then, how this Północ battle is being fought simultaneously against the forces behind 13 other proposed power plants. That's a constantly moving mountain of head splitting documentation.

And while the lawyers are dealing with that, they also take time to listen. Lawyers listened out for the locals' true concerns.

The retired teacher Mariusz Sledź minded that air quality standards would be rendered several times worse. Still more important to him was the quality of the soil. This was one of the richest areas of Poland for its soil quality. After a few years, the emissions from the power plant would fill the fields with heavy metals, including mercury. The metals would not dissolve, the quality of the crops would suffer, and they would contain radioactive substances. He knew farmers were worried they would be unable to sell their grain when it went up for its annual inspection by buyers.

'Pomerania is already high for figures of death from cancer,' Monica added. 'This will make it worse.'

Monica's mother was most concerned with the quality of air that children would have to breathe. She felt that investors were twisting environmental assessments to meet their own needs. It wasn't possible, in her mind, that such a power plant would have no impact on air quality and would make no noise.

The initial challenge involved a supplementary EIA carried out within a building permit. ClientEarth presented to the first administrative court ways in which the locals' health and interests would be affected.

'It was a huge relief,' Monica said, recalling the difference it made to have lawyers on side. 'The help we needed most was professional legal help. Legal competence, that's what the local people needed. We really felt lost because we didn't even know how to write sentences in legal terms, to make them comprehensible to the local authorities. We didn't have any idea of the administrative stages — we knew nothing.'

It was no easy triumph, however. The court accepted that the locals would be *factually* affected should the new power plant go ahead; ash would blow beyond the 200 hectares of the site, for example. However, their homes and land lay *outside* the actual site. In the legal sense linked to the building permit, that fact meant they were not *materially* affected. The court dismissed the complaint.

Let's pause for a moment. You get your day in court, you have your say, and the verdict comes back. You are *factually* but not *materially* affected. Case dismissed. How does that feel? Even if your mind lets you get inside the linguistic riddle, do you feel your actual concerns were addressed?

A primary concern for Marcin, as with all the environmental lawyers I met with in Poland, is that their work helps build civil society. I took tea with Professor Maria Kenig-Witkowska, Professor of International Law at the University of Warsaw. 'Even if there is a small thing happening in a municipality to fix our problems, there is still them and us,' she reflected. 'It is never "we are all in this together". It comes from the old Communist time, when the Party always knew what was the best for citizens. The next step is a kind of war, a battle. Maybe the next 20 years will teach us how to do it.'

'A battle' in this instance meant not giving in. Marcin and his team had taken time to listen to the locals, and helped them shape their grievances so that they might be heard. Now he had to prove to them that the courts were truly open to them, that they were not antagonistic.

ClientEarth gathered together 73 farmers and other stakeholders, and posted an appeal to the verdict in their name. Marcin argued that because the local authority had not allowed them to submit their objections in the original permitting round, their procedural rights had been broken.

An appeal was granted, to be heard in the regional court in Gdańsk in around a year's time.

The developers already had the building permit that allowed them to start work on the site. In a separate move, ClientEarth's lawyers won an injunction that stopped any building work before the appeal could be heard.

Februaries in Gdańsk are cold, but the courtroom had the warmth, the tang, the closeness of a packed sheep pen. It was so tightly packed with farmers, villagers, and townsfolk that it almost became its own environmental hazard, filled with exhaled air.

Poland has a mix of professional and elected judges, and three professional ones filled the bench of the modest Gdańsk courtroom. Lawyers for Polenergia, the developer, teamed up on the one side while Marcin Stoczkiewicz led the ClientEarth presence from the other.

'The court was attentive,' Monica's mother, Aniela, mused. This was a revelation, that local people could head into court, where they thought they would be out of their depth, and in fact receive a fair hearing. Whenever the three judges looked up, they found a whole array of citizens staring back at them. 'They listened carefully and asked detailed questions.'

The locals all had their say, and then filed home and waited for the verdict.

'A Polish court on Thursday over-turned a construction permit for a huge coal-fired power plant near the Baltic Sea, after a legal challenge from environmental groups,' Reuters reported. But, of course, victory is temporary. 'The firm behind the project said it would go ahead.'[17]

ClientEarth's lawyers switched their attention to the coal plant's EIA permit. Waters used to cool the plant would be discharged into the Vistula River. The Regulation of the Minister of the Environment, October 4, 2002, was quite clear that such discharges could not raise water temperatures more than $1.5°C$. Tests showed Północ's discharges would raise the temperature of the Vistula by $2°C$. Species of fish would die. In August 2013, ClientEarth Polska teamed up with EkoKociewie, Association Workshop for All Beings, Greenpeace Poland, and WWF Poland to submit a complaint. In February 2014, Poland's General Director of Environmental Protection ruled partial invalidity of the EIA permit.[18]

You win. You go out and win again.

On December 30, 2015, the Governor of Pomerania fully reversed the decision on the Północ construction permit. 'The governor found the permit process fraudulently limited public participation and overlooked information submitted by parties other than the investor. Polenergia must begin the permitting process again.'[19]

Before ClientEarth's legal challenge, Polenergia was vesting its future in coal. After the challenge, it announced that it was withdrawing from coal and shifting its focus to gas, the development of the electricity grid, and wind. In 2016, Polenergia put Północ up for sale. In the meantime, they appealed the Governor's decision to the court in Gdańsk.

On December 6, 2016, the judges announced their verdict. They upheld the Governor's decision. The court found that the regional authority, Starosta Tczewski, had hindered citizens' participation in the legal proceedings while discounting evidence submitted by ClientEarth and other NGOs. Dangerous emissions, such as methane, nitrogen monoxide, dioxin, and furan, had been underestimated. The building permit for the new plant, the largest new such installation planned in Europe, and set to emit over seven million tonnes of CO_2 a year, was revoked.

The small team of lawyers combined with concerned citizens had managed to turn history inside out. Rather than 2017 see smoke twisting out of the high stack of Północ, its investors were forced to return to 2011 and the initial process of applying for a building permit. This time, should they do so, they will know that lawyers are on hand to subject any application to the most intense scrutiny. They will know that the courts are prepared to keep a watchful eye on the legality of the whole process.

And the public stand alerted everyone to the dangers of a coal fired existence. At the end of 2016, the European Environment Agency released its latest report on Air Quality in Europe.[20] In terms of benzopyrene in what we breathe, a graph shows Poland to be stratospherically ahead. On maps, clusters of red and dark green dots suggest avoiding Poland unless you are intending to breathe in benzene, cadmium, arsenic, lead, and nickel. The graph for quantities of $PM_{2.5}$ in the air shows Poland once again to be way out ahead, alongside Bulgaria.

This is all accounted for in a separate table showing years of life lost per 100,000 of population. For Poles, the figure for the effects of breathing in $PM_{2.5}$ alone is 1,520 years lost. Strikingly, the other countries atop this hazardous list all come from the old Eastern Bloc — Bulgaria, Macedonia, and Kosovo even beat Poland. The lawyers of ClientEarth Polska are now bringing their attention to Central Europe. Of course, mass burning of coal blights the air beyond the countries where it is burned (those air quality tables suggest an urgent move to Iceland is the best bet in Europe), and speeds climate change, which can destroy life as we know it.

'The court's ruling affirms that local communities must have a say in such projects,' Małgorzata Smolak announced for ClientEarth. 'Along with local residents, we raised objections to this investment, but they had been ignored. We are happy that the Pomeranian Governor and the court sided with us, and ensured the rule of the law.'[21]

'Poland is not allowed by the elite to be anything other than pro-coal,' Matt Phillips of the European Climate Foundation (ECF) advised me. 'That is not necessarily where the public are; they are pro-renewables and not pro-coal. The elite are on a journey, and they will allow no opposition to it, so an overt opposition to coal was a positioning that the NGOs in Poland were very unwilling to do.'

ECF started with a grant to Bankwatch, but needed an NGO to take the lead. ClientEarth was already in position in Warsaw, and stepped forward. 'It's just brilliant, what they've achieved there,' Matt continued. 'It was wonderful to see it unfold. They built that fearlessly and they were severely attacked by the Polish government.

'Initially, there was a debate about doing the Kingsnorth thing, taking on one case spectacularly. We were of a view that that wouldn't work in Poland. What we actually needed was a comprehensive planned approach, finding a legal angle with every plant, doing whatever was necessary to disrupt the coal plants through administrative procedure interventions. The play was to stop a major wave of coal. At the turn of 2007 to 2008, we had to take on 112 new proposed coal plants across Europe. At the time, there were 20 already under construction. It's a real threat. ClientEarth were instrumental in the UK and other parts of Europe, especially Poland. In the end, only three of those projects, maybe four, have broken ground.'

I met with Connie Hedegaard when she was the EU Commissioner for Climate Action. As Minister for the Environment in her native Denmark, she experienced the strong effect of a 'green' opposition. Poland was proving an eye opener. It lacked Denmark's green opposition in both its press and its Parliament.

'Even if the NGOs really work on it and make the case, it's not in the media and it doesn't get politically dangerous,' she told me. 'The normal checks and balances in this system work with more difficulty in Poland. That's related to the fact that they don't have an opposition to whom it's important to be more green or less green. It's perfectly OK not to be interested in carbon reductions for instance.'

Work at building a civil society has a long way to go. A 2015 survey by Poland's environment ministry found that three out of four Poles limit water consumption, and 60 per cent choose cycling or public transport over cars when they can. Sixty seven per cent turn off lights when leaving a room.[22] It's a start.

I stood in fields strewn with poppies and cornflowers, and looked out across the putative home of the Północ power station. Its 200 hectare site is bounded to the west by a railway line. To the south, east, and north run local authority roads. The roads and railway count as an unnatural barrier, so the area that might be affected by the introduction of a power plant officially stops at those boundaries. You can see across the Vistula River to two forested Natura 2000 sites. The trees in one of the forests are special in that they grow on damp ground, and those in the other are special for growing in the wet. Now the Północ site's for sale, if you're interested, but my visit was back in 2013.

I stood near a farmhouse. A lady strode out. Her Croc sandals were white with a faded floral pattern, like she had been treading through old meadows. Tawny hens fluttered out of her way. A ginger haired toddler, a boy, gripped her hand. The lady's face was round and topped with close cut curls, and her eyes were light blue.

She nodded and raised a hand to point. Yes, that's where the power plant would be. She and her family would get to retain just two hectares to keep a little farming going. To her mind, all was very beautiful the way it was. She even liked the neighbouring wind turbines. They act as weather warnings. When they stop turning around, she knows a thunderstorm is coming.

The boy broke away to play on the metal spurs of the farm equipment. The lady was dressed for work, a bright blue smock over her grey T-shirt and brown skirt, but she gave me time.

She was 58, born in the mountains in the south, but she came here in 1958. Now all would change. The whole village was riven, and even families were divided among themselves. Some wanted the power plant and others did not. She explained to the neighbours that new jobs would not be for their kind — the jobs would require skills that they did not have.

Had she gone to any of the consultative meetings in town? I asked her.

No, she hadn't. What's the point? What will happen will happen, and those in opposition were much too small to stop it.

No, I could tell her now. The opposition was not too small. The locals teamed with a band of lawyers, which made them big enough to respect and to fear. They used the law. It worked. They stopped it.

How to start a public interest environmental law group

James Thornton

Not many have started up a group of lawyers to fight for the environment. One reason may be that it is not easy. In Europe, a number of smart lawyers and at least one global organisation tried before ClientEarth, and failed.

When I set up the LA office of NRDC, it was a first. No other national environmental law group had an office in Southern California. Several had tried and failed. When my Zen teacher challenged me to take care of the environment in Southern California, though, I got motivated. So I got NRDC to agree to pay my salary for a year, with the deal that if I could design the programme, recruit a top staff, and find the funding within the year, we could open the office.

The reason my gamble succeeded was the method I deployed. I spent the year in an intense set of meetings. I met with all the local and regional environmental groups. I learned their expertise, their concerns, the problems they faced. I assured them I was setting up a group to help them, not take away their territory. I learned the environmental issues in detail. As I gained expertise, I learned who the legal experts were. I met with them. The two leading ones joined the team. I met with funders. They explained that other groups wanted to raise money in LA but did not want to help it or commit to it. When I told a potential donor on the spot that I would move to LA, she wrote me a cheque for $100,000. This was rather a lot of money in those days.

Nearly everything about setting up this kind of effort is harder in Europe.

Lawsuits are harder to bring, and the laws themselves are less enforceable. There are more restrictions on charities, from accounting rules to free speech. Funding is harder to find — though there is equal wealth, fewer of the wealthy are inclined to donate. Lawyers are more likely to want to keep their work secret, even their successes, making it difficult to educate the public and funders about your work. Twenty eight different legal systems mean that, despite the centralising function of EU law, it is more difficult to act across the continent. Many Europeans, including environmental organisations, felt it was not appropriate to use law strategically for the environment.

These cultural barriers help explain why there was no pan-European environmental law group before ClientEarth. The key was adapting.

I used the same method that proved successful in California. There, I qualified for the California bar. Here, I qualified as an English lawyer. I got enough funding together to spend a year and a half learning EU law, designing a programme, finding a staff, and finding the funding. I was under considerably more pressure in the EU when it came to funding. In the US, I had a large national organisation behind me. They could help with time consuming things like accounting and payrolls and all the rest. There was a national board of trustees behind me. I had a national brand to sell. And if I'd failed, I could have gone back to other work. Here, the very concept of public interest environmental law was unknown and for many was questionable. So I had a year and a half to succeed or else fall off a professional and personal cliff.

First, I studied the law. I developed a map of what I thought the most significant problems were, and how we could use law to tackle them. With this in mind, I went out once again to an intense year of meetings. I met with everyone I could find who was an environmental lawyer on behalf of the planet. I met with executives from all the environmental organisations who would talk to me, and not all would. In California, I had learned the views and methods of the local groups, and what their needs and concerns were. Here, I did the same, concentrating on the UK, Brussels, and Poland. And as I did there, I assured groups that my intention was to supply expertise that would help empower the entire environmental movement.

With every meeting, my mental map of problems and possible solutions became clearer. My intention was to work in the EU, and then expand outward. It was apparent from the beginning that we needed a Brussels office, since Brussels is the spring from which EU environmental law flows. It was quickly apparent that we needed a Warsaw office as well.

Warsaw was key to my vision for three reasons. Poland was building a fleet of 14 huge coal fired power plants. We would need to be there to push for saner energy solutions. Then too, it has the best biodiversity left in the EU. Its Białowieża Forest is the most important mixed forest in Europe, its Baltic wetlands the most intact, and its Vistula River the last great undammed river. Finally, we needed to be in Poland because it is the most important of the new members of the EU, with the sixth largest economy and population. It is very conservative, and sends a large number of representatives to the European Parliament. If Polish politicians were to come up to the level of Western Europe in understanding environmental issues, I felt we had to be there to help them.

One needs to help one's own team too. I have long relied on meditation to help me deal with stress and gain access to creative solutions. Recently, staff members in each of the ClientEarth offices invited me to share meditation practice with them. When I started meditation more than 30 years ago, it was far from the mainstream. That has changed. Meditation is now taught in Google, Microsoft, and General Foods, and at dozens of the leading law schools in America. Judges and prosecutors are meditating, schools are teaching it to students, and there is a meditation group in the Parliament in London.

I started teaching meditation at ClientEarth in the Warsaw office, through a series of seminars. We explored the scientific basis, the physiological benefits, how to use it for stress reduction, and how it offers the opportunity to find breakthrough solutions to difficult problems. The seminars in Warsaw went deep as we meditated together and explored its connection to our lives and work.

The meditation seminars now include the Brussels and London offices, and there are plans to have weekly and perhaps even daily meditation sessions for those who want to join in. There is no doubt that the practice will help reduce stress, and that will be of great benefit. In our work, we are driven by the desire to help people and the environment, rather than the desire for personal gain. Our goals are large and the problems we address are growing. Without a methodology for coping, it is easy to burn out. Meditation offers a profound answer to this inevitable problem.

The kind of law we do is creative, and the open space that meditation makes invites creativity in. The Muses are always waiting for us to listen, and meditation opens the mind so we notice. So I am particularly interested in how meditation will open up creative intuition, giving members of the team a heightened ability to find positive solutions to seemingly intractable problems.

Earlier in this book is the example of how meditation allowed me the creative breakthrough that led to a large area of the California coast being saved from development, along with the California gnatcatcher. That is a practical example of how creative breakthroughs come. The recent awards from the *Financial Times* as the most innovative lawyers in Europe, mentioned in the opening of the book, are an objective record of how creativity is flourishing at ClientEarth. My own work in designing and building, first, the California office of NRDC and, for the last ten years, ClientEarth, depends on the calm and the insight that meditation gives me. As the effects spread out, I am sure many benefits will emerge in a nonlinear, that is, creative, way.

8

The Forests of Africa

The third floor of the law department of the Ghana Institute of Management and Public Administration has a balcony that overlooks a forest reserve. I stood and looked out. This was a lull in a brief run of days that saw taxis ferry me around Accra to interview folk who focus their working lives on preserving Ghana's forests.

At last, I had got to see a tree. The forest canopy was thick, and its borders were demarcated in sharp lines. The reserve was originally set up to maintain a green buffer between Achimota School, an elite co-educational boarding establishment, and the rest of the city; students could enter the reserve for nature lessons, and harvest it for fuel. Today, the city surrounds and encroaches on the forest. Even my vantage point brought about some shrinking, as the campus bit off a section of the forest for its own growth. Christians now strike camp between the forest's trees, 180,000 a year streaming in for open air prayer meetings. Plans are afoot to house them in their own spiritual enclave. The rest of the forest will become an eco park, where tourists can wander safari walkways and peer at animals imported to stock wildlife enclosures.[1]

It's not a bad idea. The reserve has already shrunk by 27 per cent, so any plan that retains this green remnant in a city that explodes with construction seems sound. Everybody should live within a day's hike of a tree.

I turned from the view to the buzz of Law students running up the stairs. They were young enough to still be in their growth spurts, and they bubbled in conversation as they waited outside their professor's door. A module on the environment is a core part of their Law course. The module's focus is international, which is likely the best place for beginners to start; Ghana's own environmental legislation is a scattered affair, largely hidden until recently and rapidly evolving. Shamans used to be custodians of the forests. Now it will be some of these students. Forests are being encoded in national and international laws, while timber in Ghana is now one of two things: legal or illegal.

That marks a fundamental shift. When a natural resource is placed within the rule of law, legal understanding becomes paramount for those who work to protect it. I had come to learn how environmental campaigners were managing the transition.

I had also come because there's a chance that smart injection of law into the proceedings will make a real difference. It needs to. From 1940 to 2000, Ghana lost more than 90 per cent of its trees. The World Bank's insistence on increased export revenues for the country meant exports of timber on a massive scale without any thought of sustainable harvests. Elements such as the value of biodiversity or the role of trees in regulating climate failed to enter the economic debate. Communities who lived local to the trees had no say in how their forests were treated, and gained no benefit from their destruction. All was almost too late.[2]

Ghana is still known for the high quality of its timber. In 2006, the European Union bought more than 60 per cent of all Ghana's timber exports. The global financial crisis trimmed their need for timber and forced Ghana into exploiting other markets. Still, from its position as the dominant market partner, Europe had begun the process of bringing some ethical responsibility to its purchasing power. The year 2013 brought with it the implementation of the European Union Timber Regulations to ensure timber brought into its member states was legally harvested. That would require information, including documentation of legal production from every step of the process. As an alternative, countries could enter a Voluntary Partnership Agreement and so obtain certification that all that country's timber was legally harvested. Government officials and industry representatives would develop the mechanics of bringing effective regulation into place, and, for once, civil society groups would be given a presence around the negotiation table.

Place community representatives at the heart of international trade negotiations, and you have to wonder whether anyone will ever hear them speak. Bombast, legalese, self-interest, money — there's a whole host of factors that might shout down the voices of those who stand for interests other than their own. I had seen how rural communities in Poland discovered the power of both speaking and being heard. They found the power to stop coal fired power stations being built. Might West African villagers halt the destruction of forests?

I stepped from the airport into an African night, and a familiar face was welcome. Jozef Weyns had come out to meet me. A Belgian lawyer, he travelled the globe before settling on a profession. For some weeks, he worked in a Bolivian jungle, exercising a rescued panther in the wild. At night, the roar of logging trucks broke the more subtle weave of jungle sound, and some understanding set in place. Its outcome was uncertain, but one thing was clear: shady loggers were on one side of a divide, and Jozef was on another.

Work as a development and agricultural lawyer took Jozef to Burundi. The switch to environmental law in Ghana was a natural one, since a healthy environment with equable treatment for all is also a development issue. Funding came from the UK government's international development arm. Three lawyers based in ClientEarth's London office shared responsibility for Ghana, the Republic of Congo, and Gabon. A local lawyer in each country was appointed to double the team, making six lawyers responsible for around 850,000 square kilometres.

Our taxi from the airport reached a crossroads, where pedestrians and vehicles nudged around each other like in a busy marketplace. A few lights flickered on the edges of the streets, but the traffic lights were not functioning, and so I learned my first two words of Twi — *dum* meaning off, and *sor* meaning on. Put the two together, *dumsor*, and you have 'off and on', the Ghanaian term for power cuts.

The country's timber industry has neared collapse in the last decade, with the difficulties of running major plants during constant power outages being a major factor. Protecting timber is no simple matter of battling corporations: all sides are engaged in defining what 'legal timber' actually means and how to achieve it. All sides have compounded difficulties to face.

'Be careful,' Jozef called out as the taxi broke free from the crossroads and spurted through the darkness. The taxi had no safety belts. 'My girlfriend likes my face. She will be angry with you if an accident spoils it.'

The driver laughed.

'So when will this end?' Jozef asked. The government had been promising a solution to the power crisis for months. 'When will the government stop *dumsor*? Next month?'

The driver laughed again. Jozef has an automatic mode of bringing people onside. It's a good characteristic for someone in the aid business. One thing to avoid is any move that smacks of the residue of empire, a sense that is likely acute for a Belgian mindful of a history of atrocities in Africa. W.G. Sebald recalls the writer Joseph Conrad's impression of Brussels as 'a sepulchral monument erected over a hecatomb of black bodies, and all the passers-by in the streets seemed to bear that dark Congolese secret within them'.[3] Jozef was determined not to impose any Western 'how to save the forests' template on Ghana. The objective was to empower the work of local groups with the acumen to handle the new legal frameworks placed around forestry issues.

First, Jozef and his in-country associate lawyer Clement Akapame had to make sense of the laws for themselves.

A book sits on my desk as I write. It's a hardback, its title embossed in letters of gold on dark green: *An Overview of the Legal Framework of the*

Forest and Wildlife Sector. The letters have a slightly random alignment to give the book an artisanal quality. It's one of almost 200 in print, and emerged from the Accra print shop of a former Law student. I value it as a bibliophile. Others put a keener value on it. They have given it a fresh title too. For them, it's the 'forest bible'.

Those were the people I went to Ghana to meet. They came from Accra, but flew in from the northern Highlands as well, and the forests of the west. Many carried the forest bible with them, and some carried an app with a briefing on land titles on their smartphones. They were leaders in civil society, representing different groupings within the environmental charity sector, and all met regularly as part of a ClientEarth sponsored legal working group.

What did this green book give them? Ghana has been endowed something of a mishmash of pre- and post-colonial laws, fused with customary laws that are often simply oral and rooted in tribal traditions. Some laws had lapsed, some had clauses that were superseded, and no one knew where to find a coherent sense of current legislation. Jozef and Clement dug through the shelves and blew off the dust, and the forest bible serves up the results. It is an accessible anthology of the workable laws, designed 'to help those working with forest and wildlife laws to find their way in a maze of different texts'.[4]

A choice phrase in that sentence is 'to find their way'. I had one big question on coming to Ghana: can Africa's forests be saved? And I was set to put that question to people who were giving their lives to saving those forests. Before I met them in their legal working group, I asked to speak with some of them privately. They were doing it 'their way'. What inspired them to take action? What gave them hope?

It was first thing Monday morning. Friday had dumped a whole world of problems on Samuel Mawutor's desk and he should be dealing with them. Instead, he slumped on his chair and faced me.

I turned on my recorder, checked its red light, and threw in my first question. Samuel was a programme officer for Civic Response, and coordinator of the NGO coalition Forest Watch Ghana. So what triggered him to get out there and act on behalf of the environment?

The response was electric. He was no longer weighted with civic responsibilities — instead, he was a kid whisked back to the early 90s. His eyes sparkled. 'Six thirty, Sunday evening, everyone knew what was coming next,' he said, 'and that's *Captain Planet.*'

He looked up through stylish black glasses towards the high corner of the room, as though Captain Planet might zoom in at any moment.

'We loved the storyline, the idea of a hero, and that someone was fighting against pollution.'

It was extra-special that a top US TV show featured a young kid from Ghana. Kwame was the lead among five cartoon kids from around the world. Gaia, the spirit of the planet, invested him with the power of Earth.

The show puzzled young Samuel. 'It wasn't clear to me what they were really fighting for and why Captain Planet was created in the first place. Good lessons, but how to apply them was different.'

Samuel was an urban child. The environment was more of a theoretical issue than an immediate concern. Soil erosion was very common in his area, and deep gullies would suddenly appear in the neighbourhood. He saw some roofs ripped off by sudden storms, and so for the young Samuel planting trees was more a matter of providing a windbreak than regrowing the nation's forests.

He grew up to study Political Science. That theoretical sense of the environment suddenly became political. He sensed that governments of the southern hemisphere were prioritising the needs of international corporations rather than their own people. His first boss at Civic Response, Kyeretwie Opoku, was a natural resource lawyer. Samuel had seen the need for change, and now through the model of his boss he found a tool to effect it. 'His strategies were all informed by law. Learning from him was one of the important learning points of my life.' In 2010–11, there was a constitutional review. 'Civic Response was able to mobilise a natural resource coalition. We worked with peasant farmers, Forest Watch, the gender movement and women's rights network. Because he was a lawyer, most of the work was on him: writing legal analysis, finding colleague lawyers. The memo we submitted was a very comprehensive analysis, explained articulately and referring to previous laws.'

Activist lawyers tend to burn with zeal for a certain while, and then switch focus — they have a family, need a mortgage and maybe a car. And they have qualifications that will earn them much more in the private sector. Samuel's boss, Kyeretwie Opoku, led negotiations that set up the Voluntary Partnership Agreement. He did his stint. Now he sits on the board of the Ghana National Petroleum Corporation.

His absence left a vacuum. ClientEarth moved in, not to replace one lawyer with another but to build up the legal skills and awareness of the young activists who remained.

'Previously, it was just one person who understood the law,' Samuel recalled. 'We miss his strategy and understanding of the forest sector; he's a smart person, you can't replace a smart person. But we have learned to modify our campaigns so we don't have to rely on him. Now because of all the legal training, it's not just one person, it's several. The forest bible informs us of the kind of argument we can make. Now when we argue, it is on our understanding of the law. We have a lot more pocket lawyers.'

I met this term in Ghana — pocket lawyers, regular folk who have a book of the laws tucked away in their pockets. Samuel and his colleagues wrote up the minutes from a working group and sent them to a government lawyer. 'I can see it's your lawyer who has written most of this,' the government lawyer declared. Samuel laughed. He told him their lawyer hadn't even seen it. 'Ah, then,' the government lawyer concluded, 'I see that a lawyer has trained you very well.'

The legal training has helped Forest Watch to prioritise the battles they can win. Preparation of the new Conservation of Forest Laws was previously seen as the government's job. Now that they are armed with legal training, the members of Forest Watch recognise they have important contributions of their own to make. One immediate issue stood out in which the law was clear: custodianship of trees that occur naturally outside of reserves is vested in the President. 'But it doesn't make sense,' Samuel said. So they raised the argument, again and again, until the government bought into it. They have now engaged in a consultative process to reallocate ownership of those trees and the benefits that accrue from it. They aim for a result that will make sense to local people, so that those people will work to bring it into effect.

'Most arguments we make are now pointing out violations of the law,' Samuel said. 'Our work is no longer based on value judgements, so it's weightier than it used to be. The use of law is very fundamental and strategic.'

Samuel is not a lawyer. As he spoke, I came to realise that the use of law to protect the environment is no longer the sole domain of lawyers. The responsibility is percolating down through all of civil society.

The suite of bungalow offices on the University of Ghana campus used to be decorated with the distinctive panda logo of WWF. In 2014, WWF quit the country, but its employees chose to stay together and formed NDF, the Nature and Development Foundation. Their logo is a butterfly, with its wings folded rather than spread. This new organisation plans to settle rather than fly. 'We had the option to find other fields,' Glen Asomaning, the Project Director, recalled, 'but the challenges for Ghana and West Africa are quite unique. You can't distinguish conservation from development. There's a lot of emphasis on development; everybody thinking of where to clear the next forest to put down buildings. In Accra, almost everything is gone, it's concrete, unlike southern Africa.'

He's teamed up with Mustapha Seidu to run the organisation. They are something of a double act: Glen larger and more ruminative, speaking slowly after reflection; Mustapha jumping with a quickfire energy to get out there and save the world NOW. Energy like that clearly comes with a sense of impending burnout. 'Look at implementation of Forest Laws, and you lose your energy and wonder if the little contribution you are making is building up anything,' he said, 'especially when you hear from government regulators who should know better and set the pace. I felt I would not have the energy to finish this work, because of a lot of inefficiencies and parochial interests. The real problem is implementation — there is so much to be done in terms of reform of the system itself. It takes a lot of energy and commitment from the government.'

Both Mustapha and Glen took undergraduate degrees in Natural Resources, and went on to take a Masters degree in the field. Mustapha's Masters was from Costa Rica, and a US State Department Fellowship took him to Oregon. Environmental problems are vast, and since education showed him how to understand those problems it seems he won't stop; he's adding a Law degree to the mix.

Mustapha grew up in a village in northern Ghana where water was pulled up from a well, and no one used fertiliser. The local farmers waited on dry land for the rains to come, and grew confused. By August, the maize should have a full month's growing inside of it, two months for the yams, but the farmers were still waiting to plant. Mustapha saw this, and only through education came to understand this was the result of climate change.

The forests around his village were all cut down, largely for charcoal. One patch of trees had been left standing. In 2012, it was gone. 'I asked, why did you remove it? The chief said that small bit of forest was covering us up. I realised where he was coming from, but he has no clue of the long term impact, and no way of quantifying the value.'

This speaks to a tug at the centre of forest governance: between the keen impulse to give local communities control over their own lands, and the yawning need to educate them so they know how best to act. A tree had roots, it was entirely local, but its fate likely hung on currency exchange requirements of the World Bank. Rains were delayed because of climate change, and that in part was spurred by the loss of local forest. Mustapha would drive to the city of Kumasi and pass locals who were selling endangered species by the side of the road.

'We have a law saying people should not cut down trees without authorisation,' he said. 'Well, people are used to cutting trees without any authorisation, and they will not change. Let them know the reasoning behind it and buy into the idea. There should be consequences for obeying.'

I like the term 'consequences for obeying'. It carries a whiff of menace, though it holds the same ground as another phrase I first met in an African context: 'benefit sharing'. ClientEarth lawyers in Gabon supported civil society to draft a decree to implement the vague provision in the Forest Law for the first time. The notion is that a community would not lose its forest without some recompense; a portion of money from the sale of timber would be returned in the form of a school building, perhaps, or a doctor's surgery. It is a boost for social and human rights rather than a direct one for forest preservation, but is a step in the right direction. If communities are part of the solution to better manage forests, they should reap some of its benefits.

Glen seems to have reached a level of benign pessimism. He grew up in a village in the south of Ghana, and then moved to the town of Tarkwa. The town had been mined for a hundred years, but the mines were deep and so the town was still green. A tyre manufacturing plant had a wall that bordered a forest reserve, and he did not know this was a reserve until he got to university.

Over the last ten years, Tarkwa has utterly changed. Unofficial mining has joined in with the official operations and turned the town to clay. 'It's hard to get people to accept that land you could sell for $50,000 to $100,000 should be planted with trees,' he mused. 'Government is all about working to make more revenue. We need organisations to mix conservation and development, or maybe in ten years' time we won't see any green at all.'

Some chiefs have agreed to set aside globally significant biodiversity areas. Other chiefs two kilometres away swap the timber in their forests for revenue. The disparity is too great. The set aside area gets raided illegally.

'Maybe when everything is gone, people will realise we have to protect' was Glen's final hope. 'People have been in these situations before and turned things around. California is shouting more about climate change because of drought. In parts of South Africa, all wildlife has been eaten, in Namibia too, and now you go there and see wildlife. My big concern is that we don't destroy all our resources before having to start from scratch, which would be very difficult.'

Alongside ClientEarth, NDF had done a survey of Ghana's timber industry to find out how in tune it was with the ever shifting state of legislation. NDF works with industry to try to find solutions. Mustapha meanwhile focuses on the government. 'Until we sue the government, things will not change. Sometimes they challenge you to sue them. The courts can help them by giving them an interpretation of the law.'

Mustapha was keen on some form of funding other than the 'gambling' of writing proposals to secure litigation. His Law degree would be a fresh tool he could use. 'When government and industry know you know your stuff, the negotiation is from a different stance. There is a difference between telling them something is morally wrong or more market oriented, to telling them they are in contravention of a legal provision.'

Air conditioning kept switching gears, to offer the sound effects of rainfall to the conversation. Desktops were clear, but documents massed on the floor in a corner of the office. You could dig through them to discover the tale of the WWF years. The group's new incarnation as NDF still had fieldwork at the heart of it, but a new task was to gather in information for forest audits. The legal training workshops had become fundamental to their approach.

'You have individuals calling themselves pocket lawyers,' Glen said, of the folks and their copies of the forest bible. 'Whereas people would go arguing based on sentiment, this time people are arguing based on fact, because at least they have that book which they can easily refer to. You face different people that way, and you go feeling more confident than you would otherwise be. I have made this statement and I will always repeat it: that intervention by ClientEarth I see as being one of the best interventions by any civil society group in the forestry sector. It's been very useful for a lot of us.'

African borders are permeable. Men slip into the forest reserves of northern Ghana and work deftly with their chainsaws. Seventy per cent of each felled tree is left behind, but the rest is sliced and made portable. Chainsaw logging is illegal in Ghana, so the timber is carried across the border and sold in Burkina Faso.

That same border that lets through the loggers also demarcates where sub-Saharan desert begins to spread, after a run of grassland. On the Burkina side, the savannah is still replete with trees. It is in the Bongo District on the Ghanaian side that the land is turning to desert. Tree felling without replacement is a big part of the story, linked to intensive farming practices and land excavation for road building and construction.[5]

Elvis Kuudaar has walked both sides of the border to suffer the contrast. He saw how Burkina had stopped the encroachment of the Sahara, while the desert was eating into Ghana. The loss of savannah feels like a personal degradation. He grew up in a rural community close to the border. His uncles and other farmers complained about the rains not starting. When it rained, they would go out into the fields to sow their seeds, and then the rains would stop. It was hard for them to predict.

Elvis too studied natural resources management at university. He began to be able to explain to the farmers what they were going through. When he went to work for the Environmental Protection Agency, they based him in the east of the country, where the Volta follows its course. Variability in weather patterns became a major problem. 'The rainy season used to spread over six to eight months,' he told me. 'Now it lasts one or two months, and the quantity of the rainfall is equal to when it was spread. There is a lot of flooding that washes away people's crop. Farmers in the dry season are mainly farming along riverbanks to make up for a shortfall in food production.'

He works as an independent consultant to various bodies in the charitable sector, and was introduced to me as the lone embodiment of a new acronym, an NGI. It stands for non-governmental individual. Language is in flight from common sense with such coinages, but Elvis does suit that non-aligned status. Just as he flits across international borders, making observations, finding contrasts, he moves between different areas of environmental concern. When ClientEarth first formed its legal working group, he moved straight in on the process, and had attended every single meeting. I met him on the eve of the latest working group session. Folk fly and drive in from all across the country, from the community leader in the west to the policeman in the north, who has taken the women who harvest savannah trees as his private concern.

Elvis drove his car to the meeting with a newfound confidence from all the legal training. 'I developed the interest of reading other laws and using them,' he told me, 'even road traffic regulations so a policeman can't intimidate me anymore. I can say, "What offence have I committed, under which laws?"'

Alongside delivering a working digest of the relevant laws, Jozef Weyns and his team devised a framework which community and civil society organisations could use to ensure that evolving legislation met their needs. The framework consisted of five 'focus rights': Ownership and Use Rights; Benefit Sharing; Access to Information; Participation; Access to Justice.[6] In Ghana, Elvis took part in the group working on Access to Justice. 'It exposes the direct reality of what impacts forest management in Ghana,' he told me. And as he spoke on, I came to see how Ghanaian citizens had felt emboldened in the same way as those townsfolk in Poland who fought to stop a coal fired power plant.

'In the early days of advocacy, we could make a lot of noise in front of government, but when we sat at the table to speak and they sat there with their lawyers, we were found wanting because they were quoting laws and sections and things that we were not familiar with. When ClientEarth began its training programmes, I started to have a clear appreciation of the relevance of the focus rights. They apply to governments, communities, timber companies — you can apply them to yourself and find a legal basis for how a legal process can change or be reviewed.

'Along the line of the training, I can now sit down and argue with state officials and be able to articulate my position better from the basis of the laws, the policies, and the policy intentions and what we are getting out today. In Ghana, we have scattered laws; some are repealed, with only one section still applying. We as non-lawyers cannot do that analysis, but legal briefs from ClientEarth, and the focus rights, have reduced my interest in studying law. I don't need it. I can read different laws and see how they relate and how one impacts on the other. Nobody can walk over me anymore.'

Alliance Française sits on acres of downtown Accra. The dining area is set under a thatched roof that is raised on beams, the cool air of an August evening breezing through. A cocktail waiter attended the bar, but had little business. A cocktail probably costs a day's wages, a pizza just a little more.

Our party ordered. Kwame Mensah was among those who had arrived early for the next day's training. The restaurant put his tales of childhood in context. Popular food when he was a youngster was bushmeat: a roast native guineafowl perhaps, or some tender slices of grasscutter (cane rats some 60 centimetres long that accrue some tasty flesh when conditions are right). In those years, this was free food. Now the supply of bushmeat has gone, with the loss of forests and climate change drying up the creature's main water supplies. Some farmers are diversifying and working to domesticate the animals; bushmeat is becoming an expensive delicacy.

Kwame used to trail farmers to work. His youth's not long past, but times were different then: farmers prepared the land in March and April, and sowed in May. Now the rains don't come till June.

Knowing what to do when: it's an art you have to keep learning. Kwame heads Kasa Initiative, a platform for a number of natural resources and environmental organisations. One aspect of the ClientEarth legal training proved particularly helpful. 'It's difficult to structure an advocacy campaign,' he explained. 'You tend to do advocacy that might not even be on the table of the parliamentarians. The legal working group came at a time when Ghana was reviewing and developing its laws to regulate illegal logging, and putting in a holistic legal framework to help government to manage its forest resources sustainably. The training provided information about what parliament does, and that helped us to target this process. Now we hit on the issues they are discussing at each sitting so you don't waste those resources.'

'We don't care about the bigger parliament,' Elvis joined in. 'We target the select committee. This time, we are not just pushing them on a speculative basis but on a concrete basis, based on law and facts. As an example, there's an off reserve legal initiative the government has been trying to pass. It's not been able to move beyond the select committee level for a long time now, because we have been able to engage with the select committee properly, raise the issues, the legal basis, including the Voluntary Partnership Agreement's interest in protecting communities and small enterprises, and the traditional forest and wildlife policy. The initiative's now going to support two or three other government intentions. Our new legal understanding lets us put our case properly, interface in the process, and engage the right stakeholders so they appreciate what our stance is. We've moved away from our noise making days and into engaging the constructive process.'

The ladies in our party collect their meals in containers, to take home and eat and maybe share later. Kwame summons the bill, and Jozef neatly takes it on board. Elvis has a tale to close the day. He was in the city of Kumasi, leading forestry commission officials through the legal process of off reserve logging. The head of operations, Dr Kuchamayo Quachamayo, came to speak to him afterwards. 'How did you learn all these things?' he asked. 'You could be a resource to train our staff on the legal process. I can tell you most of our staff don't know much about the legal processes.'

Elvis wiped his hands on his napkin and set it on the table. 'People are beginning to appreciate civil society activists now, because of the knowledge we have over legal issues,' he said. 'The dynamics are changing.'

Thirteen egrets lined the top of a wall at the edge of the hotel, alert to any flotsam in the wide drainage ditch below. The ditch water was brown and green, with dashes of deep unwaterlike blue — some mixture of wastewater, sewage, and chemical effluent. I traced the stream where it crossed between a road and down to a waterfall. A man with a spade turned the waste that was dumped by the water's side, looking for pickings. More egrets waited on the bank beyond.

Further along the road, a large government billboard announced the penalties for illegal dumping. The waste ground beside it was a metre high with illegally dumped garbage, some bagged and some just strewn. Nana Tawiah had come to the hotel to help run the day of legal training on forestry issues. This brief walk set that day in context; the environment is challenged in every area.

Nana Tawiah Okyir took his Masters in Law from Harvard and now lectures in Law, but after graduating he took a government job. 'I worked five years in Parliament as clerk to the committee on local government and rural environment,' he told me. 'We visited the major waste processing plant in the country on field trips. Nothing seemed to be happening year on year, because of a breakdown of the environmental rules in the country. Nobody seemed to care. There was a problem, but everyone had thrown their hands in the air in despair. The major lagoon that carries waste material is totally choked, which causes floods in the capital. In the '80s, there were lots of trees where you lived, even in the urban areas. They all seem to have gone. That gave an awareness of the problem.'

The walk included another university campus, this one left over from the colonial era, and the grand mixture of palace and fortress that was the Temple of ECK, the base of the Eckankar religion. On the edge of Accra, the rest was low housing, some slums on a hillside ranged around pockets of market garden, and the occasional stumpy tree. More construction will soon sweep the gardens, trees, and temporary housing aside.

It was time to head back to the start of session. 'In the legal group, people have experiential knowledge that goes beyond the law,' Nana Tawiah explained, 'which helps you understand the law cannot solve that problem, because it is cultural. For example, the forest official who lives in the area won't arrest a hunter in the close season, because the people rely on bushmeat. It's not realistic. You can pass that rule in Accra, but the person in the forest appreciates this is what people live on. They won't implement it.'

Nana Tawiah and Clement Akapame, who fronted ClientEarth's operation in Ghana, have started up a new not for profit law firm and named it after a combination of their mothers' family names, TaylorCrabbe. They hope to establish themselves with victory in a significant piece of litigation. Interestingly, that might mean taking the issue of special permits to court. These are an anomaly in forest decision making, through which the minister might grant a logging permit even though the normal safeguards and guarantees are not met. The lawyer who headed Civic Response initially argued special permits out of the Voluntary Partnership Agreement, excluding them as a legal source of timber, but they kept creeping back in. Time and again, leaders from civil society groups told me how their legal training had empowered them to argue the point all the way to the minister's office. They came away with a sense of victory, but it is unclear whether in the end the minister still issued special permits when it suited him. Reasoned argument can only take you so far. It becomes toothless without the capacity to enter the courtroom.

The Erata Hotel has a pleasingly '60s feel to it: the sharp lines of its sloping concrete are painted sky blue, and shaded corridors allow a natural flow of air and open out onto a courtyard bar with tables arranged on layers above twin pools. Its conference room, on the other hand, is entirely functional. Blinds closed off any view while participants crowded round the edges of the square of tables packed inside.

Nana Tawiah stationed himself at the back, fingers at his laptop, his dark suit and tie giving him an air of presiding authority. Some of his young students sat to either side of him, keen to witness an episode in this quiet revolution of the spread of law. They fetched more seats as extra participants squeezed their way inside. In the end, those participants numbered 32: 24 men and eight women.

Jozef kicked off with a presentation on Social Responsibility Agreements — different ways in which communities get to share the benefits of logging. He was barely two minutes in before the first question interrupted him, and he had no moment to answer before a response was hazarded from elsewhere in the room. This was a legal training session, none of those in the room were lawyers, and yet they clearly had the confidence to thrust themselves into legal debate.

'We have beautiful laws,' Mustapha had told me, while lamenting their lack of implementation. One such law was the Forest and Wildlife Policy of 1994, which reformed the allocation of timber rights. Jozef led the group through the details of several aspects of these, including Timber Utilisation Permits. These are meant to allow community groups to cut a specified number of trees for their own social uses, such as construction of a school building. A study from 2011 showed 124 such permits had been issued, and yet none of them had gone to community groups. They were all assigned to timber firms. And rather than be for a specified number of trees, the average tract size for each permit was 31.7 square kilometres.[7]

Still, you can't work to implement laws until you know them. Jozef stepped aside from his Powerpoint and wielded a marker pen on a blank page of a flipchart. With a few strokes, he'd drawn a tree. A few more, and the tree was rich with foliage, stood in grassland, and had a snail crawling up its trunk. A few circular dashes, and the tree bore fruit.

The drawing illustrated the fact that a tree is much more than its timber. One savannah tree had particularly struck me in this regard. The shea tree is indigenous to a belt that runs across sub-Saharan Africa. A tree takes seven years before it first fruits, and it can be another 20 before it is at full strength, after which it can flourish for 200 years. In northern Ghana, parklands are cultivated areas where certain trees are allowed to grow because of their value to the community, and 80 per cent of these trees are shea. The fruits resemble large plums. With six months of ripening, the thin pulp layer can go from tart to sweet, and shields the large nut inside. The harvesting of these nuts is exclusive to women and their children, and provides between 20 and 100 per cent of rural female income.

A single tree has a life that will support multiple generations. Evidence of the trade in Ghana's shea butter goes back to the 13th century, with women selecting trees for particular qualities of oil through all that time. Local rules are in place to prevent nut collection when rains start to fall, because women need to prepare fields for sowing at that time. This saves intruders raiding for the spoils.[8] A mass of such evidence shows the careful husbandry of trees by local communities. It is an economic matter, but also a profoundly cultural one. Shea butter is used for cooking, to condition hair, to moisturise skin in dry winds, to heal wounds. Husks are used to decorate housing, and the residue of the butter process strengthens the houses' clay bricks. Branches garner doorways to protect newborns from evil spirits, while the shea butter soothes their umbilical cords. A drink is made from the bark that is said to help diabetes, while the leaves are used against headaches.[9]

Visual grace notes of every snarled traffic intersection in Accra are the women sheathed in radiant cloth who are so poised as they walk, with voluminous loads balanced on their heads. In the rural north, it is standard for women to walk up to 15 kilometres with 25 kilograms of shea nuts on their heads. They have an intense understanding of the trees, and work hard from their childhoods to tie the trees' cycles of life to their own.

Sadly, as forests come under threat, traditional savannah trees are targeted. Most trees are felled for charcoal; others are removed to make way for mining operations. ClientEarth was asked for a legal briefing on women's rights to trees they had harvested for generations. Modern laws had failed to take the women into account.[10] Hope was now invested in the process of constitutional reform and changes to forestry laws.

In the spring of 2014, five working parties from the legal training group were assembled and set a task: prepare a position paper on an issue of environmental concern. Those papers were now complete. The plan was to release each position paper as the process of legislative and regulatory reform in that area kicked in. On the matter of tree tenure and the specific issue of the shea trees, now was the time. One idea in the paper was to classify the shea tree as an economic plant, in the same category as cocoa plants, and bring it under the Economic Plant Protection Act. The paper would be sent to the Ministry of Food and Agriculture, the Ministry of Land and Natural Resources, and the Parliamentary Select Committee.

One man sat particularly alert as this decision was taken. Corporal Samuel Naawerebagr worked as a police detective, and had also taken the issue of the shea trees to heart. As a kid, he picked the shea nuts for his mother, ate them for his lunch, rubbed the cream on his legs, and used the oil for cooking. How did it feel to be sending the position paper as a letter to the authorities?

His face had been passive, but now he smiled. 'I started it, so getting this action, getting the letter and paper out there, is so much encouraging,' he said. 'I see some hope at the end of the day.'

The day's training was intense. Sessions ended with a quick Yes and No test to check on understanding, and again there was no diffidence. Answers were called out from all around the room. Participants were a radically mixed bunch: an indigenous priest, activists from Friends of the Earth, a police detective, an artisanal miller, an officer from the forestry commission, some esteemed members of the old guard who had been environmental activists for decades — a whole spectrum, with different interests and ways of working, who managed to achieve a common purpose for this series of training days. 'Can't you speak English?' one shouted out, as someone lapsed into convoluted legalese. This bunch could turn mean when it wanted to.

I sat at the back and scribbled down my notes, like a ghost at somebody else's conference. Those around me were engaged learners;

they posed their questions when points were unclear, and they even took turns in presenting. These were campaigners rather than lawyers, but they had taken on a lawyer's demeanour: they gathered facts, studied provisions, weighed counter arguments, as though intellectual pursuit were a game in itself. And then the day closed with an open floor session, in which folk had the chance to bring live concerns to the forum.

The issue that finally caught fire was the one of chainsaw logging. Bands of men wander into the forests to cut selected trees. It's a deft business: I met campaigners who admired the skills of converting trees into planks even while condemning the practice. Chainsaw timber has been illegal in Ghana since 1998, and yet it still supplies most of the domestic market. The writer Fred Pearce has been eloquent in defence of the loggers: 'These operators are not angels,' he admitted. 'But they are mostly meeting local needs through selective logging on existing farmland, while providing income for local farmers and employment for local communities. They are as essential to a country like Ghana as smallholder farmers.'[11]

In the same article, Fred Pearce tells how $750 per truck per day is paid out in bribes to the police to allow this chainsaw timber to pass from one region into the timber markets of Accra. Traffic patrollers were taking their slice of $100,000 per week. Empathy for individual loggers is understandable, but those individuals are serving a massive cartel that buys off the regulators. They rip apart forests so that those few who can afford the trucks and the bribes rake in personal fortunes. That is the backdrop to the upset that erupted in the room.

One man started it. His organisation worked to support chainsaw loggers in a switch to artisanal milling, where a reasonably portable machine would replace the slicing of trees into timber by chainsaw. He told of encountering chainsaw gangs cutting sections of the forest and carting them over the border to Burkina Faso.

'People are colluding,' said one raised voice, 'on behalf of massive illegal chainsawing. We're here discussing, but by the time we're finished there will be nothing.'

The whole problem was political, another man suggested. Someone would be arrested, they would be sent to Accra, they might get as far as the courts, and then they would walk free.

'We are just cutting our forest and sending it to Burkina Faso, and it is all aided by politicians,' yet another voice added.

Calls came from around the room:

'It is not a rumour.'

'Evidence is there.'

'How can we help the forestry commission, and announce that the regulators are hopeless?'

'Let's get evidence of the politicians involved!'

'Let's use the strength we have in here to do something about it!'

They would stir up media interest, and guide journalists to the scenes of destructive logging, they declared.

'We'll deal with this chainsaw thing!'

Passions were enflamed and rousing. Suddenly, a roomful of charity employees felt primed to hold a ruling elite to account.

Then the anger ebbed a little, and folk moved soberly to the next agenda item. I got stuck behind. The flurry of outrage was not specifically against the loggers, but against those in power who subvert the regulatory system. It raised the most contentious of issues which has attached itself to Africa: how is it possible to act for the good in the face of corruption?

Render a common activity illegal, as in the outlawing of chainsaw timber in 1998, and you create great potential for corruption. What was once legal is now illegal, and those who continue in old ways are now viewed as criminals. And as such they must act as criminals. Sooner than pay fines, they pay bribes, and so the regulators and politicians who accept this money make criminals of themselves. Money and power therefore become invested in ensuring that illegal trade flourishes. Jens Lund and his fellow authors of an influential paper are clear on the problem: 'the governance regime has served the entrenched and political elite in the exploitation of timber in Ghana', they attest. 'This elite has subsequently and with considerable success resisted any attempts at reforms that could threaten its favourable position.' They proceed to suggest 'that timber rights are allocated in exchange for payments and/or political support, e.g., in connection with election campaigns'.[12]

I heard the same tale from civil society leaders. One told me of a meeting they had with a minister outside of an election period. They felt listened to. Then a new minister took up the role, and an election was in the offing. 'He asked us there to see us talk, not hear what we said. In his conclusion, he disregarded everything we said. He was there to help industry.'

Ghana might be termed an extractive economic institution, which the economic thinkers Daron Acemoglu and James A. Robinson define as 'designed to extract incomes and wealth from one subset of society to benefit a different subset'.[13] Jared Diamond glosses those subsets as being 'the masses' and 'the governing elite'. Diamond goes on to set this in a colonial context, where 'Europeans introduced corrupt "extractive" economic institutions, such as forced labor and confiscation of produce, to drain wealth and labor from the natives.' Alternatively, where native populations were sparse:

> European settlers had to work themselves and developed institutional incentives rewarding work. When the former colonies achieved independence, they variously inherited either the extractive institutions that coerced the masses to produce wealth for dictators and the elite, or else institutions by which the government shared power and gave people incentives to pursue. The extractive institutions retarded economic development, but incentivizing institutions promoted it.[14]

That is the struggle in Ghana and elsewhere in Africa: to throw off the poisoned colonial inheritance and become an incentivising institution. To achieve sustainable forestry, those incentives need to go to those communities who live on the fringes of the forests and the women who work the trees of the savannah.

It's not an impossible turnaround. In Ghana, my attention was focused back to the UK, where many communities are totally opposed to the introduction of fracking in their neighbourhoods but are given no voice in negotiations. The process seemed shoddy compared to the Ghanaian model, where civil society and communities are embraced within the negotiation model that has set up and is implementing the Voluntary Partnership Agreement. While Ghana was the first country to sign up to the process of formalising such an agreement, Indonesia has made faster running. A couple of the attendees from the Ghanaian legal working group were selected to fly out to Indonesia and meet with similar group members there.

Indonesia offers an encouraging model (though I am aware how rapidly good models can collapse). The 2015 United Nations survey of forest cover reported a 50 per cent reduction in the rate of global forest loss. Indonesia was second in the table of forest reduction, but third in the table of top ten countries with forest in preserved areas. By the next report, it is expected to be reporting forest gain.[15] The election of President Joko Widodo with the primary intention to crack down on deforestation and peatland destruction is a major factor in the change. 'Community management is usually environmentally friendly,' he has stated, 'but if it's given to companies it is turned into monocultures like acacia and oil palm.'[16]

When Jozef introduced my presence to the group, they learned that I was writing this book — and returned their attention to Jozef, in the hope of information they might actually have some use for. One man was different. He had close cropped silver hair, a purple batik shirt swirled with white lines, and a sense of the value of story. He looked at me, nodded, and made sure that we met.

His name was Osofo Kwasi Dankana Quarm; 'Osofo' was an honorific, the equivalent of 'Reverend', for he was an indigenous priest.

He had travelled a long way, from the Wasa Amenfi traditional district in the west of the country. Accra was familiar to him, for he had been a businessman in the city till 2000, but then he quit. Now he presented himself as 'an environmental priest', and worked as a landowner and farmer. He was also a founder of a regional branch of CREMA, and took a few minutes out of one of the sessions to write an explanation of this for me, in neat script on a sheet of paper:

> The Community Resource Management Area (CREMA) framework is a new environment and natural resource governance concept introduced by the government that allows stakeholders within communities to be mobilized, organized and administered as a corporate entity for the purpose of pursuing their own development based on their ideas, institutions, values, customs and resources in partnership with others from the local, national and international levels guided by principles of free, prior and informed consent.

He was developing businesses out of the forest, looking for ways to recover the commercial value of medicinal plants, timber, animals, flowers, and so on. The largest timber company in the country was in his area, but he was busy planting trees so the community could build factories to process their own timber. His aim was sustainability: maintaining the environment so his community could continue to live there. 'Quality of life depends on the extent we have access to natural resources,' he assured me.

At which point, I dared to ask a question that had been troubling me for days: didn't he tire of the term 'natural resources'? It implied that the natural world had no intrinsic value beyond any material benefits it could bring to humans.

His eyes took on an extra glow and he smiled. 'In African religion, humans don't own the world,' he told me. 'We own the world alongside the animals and plants, the living and the dead, the yet to be born. That's why we don't cut down trees unnecessarily. When others come, who cut the trees, it seems a matter of law. The politicians are failing us.

'Here we don't sell land. Everybody has the right to see the chief and to farm land. It belongs to the community. Strangers intruded with the concept of land ownership. It's a big problem for the community to understand that land is in the hands of someone not under my authority. That I can't hold my festival.'

The indigenous community was shrinking, he explained, and the farming system was disappearing, for nothing was supporting it. There was no longer any surface water, and suddenly they needed to apply high doses of fertiliser to the land. This was expensive. Farming was becoming more knowledge based, and nobody was coming to the farmers' aid with the appropriate technical knowhow.

While he represented a traditional society, he was alert to the need for change. 'Empowerment comes from the preparedness of people to engage with the conditions of life,' he told me. That's why he made these trips to the capital, and learned what he could. 'If I don't carry the message to my people, then there's a vacuum.'

He was a leader in the World Bank initiative that was numbingly titled the 'Dedicated Grant Mechanism', which worked to bring indigenous communities into global discussions on climate change. And his view on law was pragmatic: 'Law is not the only thing, but it is an important step forward to have a common way of doing things.'

The day closed, and Jozef loaded a car with equipment. It had been a long and arduous day. Others had gone home, but Osofo Kwasi had stayed busy somehow. Now he breezed across the hotel's parking lot, like some gust of benediction dressed in purple batik. He took Jozef's hand in his.

'Thanks for coming to support us,' he said. He squeezed Jozef's hand, let it go, and then stepped back and smiled. 'The onus is now on us.'

Making laws work: implementation

James Thornton

When a law is passed, it starts life in the world. Every environmental law is designed to achieve a real world result. It has a targeted community whose behaviour it regulates. It has a body charged with making it work.

A law does not implement itself. It takes a great deal of work. When you are implementing a law, you are interpreting how it applies in the real world, no matter what it says on paper. How a law is implemented determines whether it achieves its original purpose or not. Industry spends a great deal of money on lobbyists who try to capture the mind of the regulator, and shift the interpretation in their favour.

As we've just seen, the African lawyers we work with are focused on implementing the Voluntary Partnership Agreements their countries have entered into to sell timber in the EU. These agreements require the strengthening of the basic architecture of civil society, such as citizen participation. That the African lawyers are so focused on implanting these key provisions that help build the rule of law is extremely encouraging.

When I was starting ClientEarth in 2007, I travelled around meeting lawyers and NGO executives in Europe. One of the things I most wanted to know about was their experience in implementing laws. As a lawyer, I know that you can win in Parliament, but lose the implementation game.

I discovered that the European NGOs focused on campaigning. They are good at it. But they did not get involved in implementing laws. Once a law was on the books, they moved on to other issues.

When I met with Tony Long, the founder of the WWF office in Brussels, he was clear. He said that all the major environmental groups in Brussels had worked on the EU chemicals law that came into effect the year before. The law is called REACH, an acronym for a regulation with the lugubrious name Registration, Evaluation, Authorisation, and Restriction of Chemicals. Tony said that, altogether, around ten NGO people from various groups had campaigned to make REACH a good law. This effort was successful, because the law, at over 800 pages, was the best globally for its intended regulation of dangerous chemicals.

But, he said, he was frustrated that every NGO working on REACH moved their people off it as soon as it was made law. Not a single full time person in Brussels was working on implementing REACH. Tony said he knew that laws did not implement themselves, and the fight needed to be continued to make the law work. This was a most frustrating example, he said, of the way European NGOs worked, but he thought it a typical example.

I was astounded. This meeting was a turning point for me. Sitting in Tony's office in Brussels, I came to two conclusions. First, I would make implementation central to the work of ClientEarth. Second, I would start a project designed to implement REACH — in the Brussels office which we did not yet have.

We did have a Brussels presence and then an office soon after. One of the first projects was to implement REACH. The work on this chemicals law can serve as a brief case study on implementation.

Let me start with an observation. There is a good reason more people do not do this kind of work. It is technical, time consuming, and difficult. Because results are incremental, both you and your funder need patience. Without the participation of the public, though, industry moves the game inexorably in its direction.

Let me give you an example of what goes on behind closed doors. A major ambition of REACH is to remove dangerous chemicals from commerce so that people and the environment are not harmed by them. REACH set up an agency, the European Chemicals Agency (ECHA), charged with implementing the law.

Central to discovering whether a chemical might be dangerous are scientific studies. The results of studies go into a dossier on each chemical. Only if the information in the dossier substantiates a risk can a chemical like an endocrine disruptor be appropriately regulated. What goes into the dossier is therefore crucial. Control what goes in there and you control the results. So how good are the dossiers?

The dossiers should contain an accurate reflection of all the information that is relevant. The reporting of the test results should be reliable, and the weight of the evidence should be adequate to support action if the substance threatens harm. We did a study to see if the system was working, choosing five endocrine disrupting chemicals.[1] Endocrine disruptors can cause cancer and birth defects. We looked at the scientific peer reviewed literature and compared it to what was in the dossiers.

The study revealed that much of the relevant information was not in the dossiers at all. In other cases, summaries downgraded the seriousness of findings. In many cases, it was not possible to compare the summary with the original study, because the summary was of poor quality and gave no information as to the author and no citation.

ECHA prefers information generated by industry using Good Laboratory Practice. But such methods do not guarantee good science, they only avoid fraud in tests. Experimental science has moved on, and peer reviewed studies based on 21st century science are more sensitive in showing whether a chemical may cause cancer, birth defects, and so on.

ECHA set up a system that will simply not reflect the best science available, and therefore not show that many chemicals may be dangerous. It therefore comes as no surprise that since the law was passed in 2006, few chemicals have yet been removed from the market under REACH.

So what can one do? I hired an Italian lawyer, Vito Buonsante, who had experience on REACH from working in a consulting firm. From Brussels, we have been systematically following the process the agency goes through, arguing for better data and more transparency, sitting on committees, and being present in the process, just like industry, and able to make legal challenges to improve regulatory behaviour.

We have also been demanding that much more information about toxics be made public. For example, the European Food Safety Authority (EFSA) regulates pesticides. The agency makes decisions through committees of experts, and many of the experts get their expertise by working for the chemical companies. The possibility of conflict of interest is apparent.

We obtained decision documents by EFSA committees, with comments by committee members. But there was no disclosure of which expert made what comment. We therefore asked to know what comments were made by which members, so we could know what paid representatives of the chemical industry were saying.

EFSA refused to disclose, saying it would invade the privacy of the experts. We argued, through Pierre Kirch of Paul Hastings in Paris, that when an expert contributes to a decision by a public regulatory agency, there is no privacy interest as important as the public right to know.

The European Commission came in on the side of EFSA, demanding secrecy. The lower court denied us access, so we appealed. The European Court of Justice decided in our favour and ordered that the identity of the experts must be made public.[2] This is one small but important step at stripping away secrecy and undue influence.

There is much to do to make REACH work, and to make ECHA and EFSA transparent and accountable, and make sure they implement the law correctly. Other organisations have joined the work on REACH, and there is now a good network of NGOs working on toxics.

Citizens are at the table, to balance the influence of the chemical companies. Without public scrutiny and input, there is no chance the law will serve its intended function. But with public scrutiny and expertise, there is everything to play for.

9

The Dragon Awakes

Beijing, 1984, and a bicycle came with my job. Few bikes had lights, and it was the same with the city's streets. I massed with others in the night-times. We waited by level crossings as trains rolled past, container after container from far away, and then we pedalled on.

In medieval alleys beside the Great Wall, locals cooked on charcoal braziers. I took a steam train across the country and looked down on patchwork fields filled with peasant workers. Men held ropes to pull a plough on which another man stood, guiding the blades into the earth beneath his feet.

That year was the 35th anniversary of the founding of the People's Republic of China. I was part of a World Bank funded initiative to train students in English so they could handle postgraduate study in the West. I sat at a banqueting table in the Great Hall of the People. The radical premier Zhao Ziyang envisioned the future in a speech from the stage, and a symphony orchestra offered incidental music from far to the rear.

In the daytime, I sat in a stand right in front of the gates of the Forbidden City. Close on my right was the Paramount Leader, Deng Xiaoping. Half a million people filled Tiananmen Square in front of us in synchronised dance. That night, the sky blazed with fireworks. We looked up and marvelled like children, till flecks of hot soot fell and burnt our eyes.

That's not so many years ago. Trains kept rolling. They brought materials for development and peasants from the fields. Cars pushed bikes into history and filled new multi-lane highways. Buildings rose and shone light through plate glass and into the night sky. And as I write this, Beijing has issued a 'red alert' for smog. Schoolchildren are to be kept at home, cars held back from the streets, outdoor construction halted, and polluting factories shut down.

Beijing closed the last of four inner city coal fired power plants in 2016, and replaced them with gas fired stations. The city, whose smog hit world headlines in the run up to the Beijing Olympics of 2008, plans remedies while the rest of the country maintains coal fired momentum. The first nine months of 2015 saw permits given to 155 new coal plants, with a combined capacity of 123 gigawatts. It 'raises questions about whether China is weaning itself off coal as quickly as it can', Edward Wong remarked in *The New York Times*, 'and whether officials are sufficiently supporting nonfossil fuel sources over coal, which is championed by some state-owned enterprises. China is the biggest emitter of greenhouse gases in the world and the main driver of climate change, and has some of the worst air pollution.'[1]

When we started this book, I met with some despair in Europe. 'What's the point in our stopping coal,' ran the general line, 'while China builds a new coal plant every week?' It felt like wiping up a puddle while a river burst its banks.

Stop coal in Europe and you not only mitigate global warming, you model a way forward for others. Such a model is redundant if nobody notices. Can China learn anything from the west? Can a country so invested in the rule of the party give expression to the rule of law? Is there any place at all for civil society in China?

I imagined negative answers to such questions. Environmental law has a way of astounding, however. Remember that phenomenal wealth of environmental legislation that sprouted during the Nixon administration in the America of the 1970s. Have hope. A remarkable story is unfolding in China.

Let me share a trio of observations from history, for context. Here's the first.

In Poland, lawyers told me of their hopes for the establishment of a civil service corps in Poland. They felt the lack of such a professional body in government, one which could carry core competencies between administrations. Currently, they felt, the elected party has so many favours to return that it needs the civil service posts for its supporters.

'By contrast,' Henry Kissinger noted in his survey of China, 'when it entered the modern period, China had for well over a thousand years a fully formed imperial bureaucracy recruited by professional examination, permeating and regulating all aspects of the economy and society.'[2] Kissinger reflected on how Mao Zedong took explicit guidance for his revolution from imperial history. 'China is singular,' he concluded. 'No other country can claim so long a continuous civilization, or such an intimate link to its ancient past and classical principles of strategy and statesmanship.'[3]

Millennia of central rule include centuries of violent upheaval and fracturing, but for purposes of affecting environmental change, systematised autocratic rule has enormous powers of delivery. This can be for good or ill. In 1959, ten years into the Republic, China endured what the State euphemistically terms 'Three Years of Natural Disasters' — which historians ascribe to state policies. Estimates for the numbers of consequent deaths range between 20 million and 45 million.[4]

The rapid development of the late 20th century has caused environmental blight. But China has a system of governance in place that can command profound corrective action.

Such corrective action amounts to a restoration of balance between mankind and nature. My second observation regards such balance. In his reflections on China's imperial age, Kissinger noted how 'the Emperor was perceived as the linchpin of the "Great Harmony" of all things great and small'. Signs of harmony's collapse, including any sequence of natural catastrophes, presaged a shift of authority. 'The existing dynasty would be seen to have lost the "Mandate of Heaven" by which it possessed the right to govern: rebellions would break out, and a new dynasty would restore the Great Harmony of the universe.'[5]

It's simplistic to equate 'Emperor' with 'General Secretary of the Communist Party', but it is clarifying to accept that disharmony causes disquiet and leads to unrest. In 2015, 96 per cent of respondents to a broad survey of the Chinese population believed their standard of living was better than a generation earlier. This is not a time of popular unrest in China. It is, however, a period when large groups cohere around what they see as 'very big problems': 44 per cent cite corrupt officials as such a very big problem; 35 per cent, air pollution; 34 per cent, water pollution; and 32 per cent, issues of food safety.[6]

In her study of the internet's effects on Chinese society, Ying Zhu explores the nature of an effective 'critical society' in China, and sees that its size needs to be 'just enough to achieve some certain effect in some particular situation'. A newly digitised population forms such a critical society, has a forum independent of state control, and finds ways to bypass government censorship. 'Vigorous public discussion and networking on the internet and via social media have become central features of contemporary Chinese society.' Government surveillance of this public dialogue alerts them to significant concerns that need addressing. Ying Zhu concludes her argument: 'In a nutshell, authoritarian states that adapt and accommodate in anticipation of a revolution might persist indefinitely.'[7]

People are disturbed by corruption and pollution. Those are the elements causing disharmony. Alert to such disquiet, it is in the Chinese government's interest to prioritise corruption and the environment as problems to be addressed immediately.

My third observation from history: the Chinese government is focused on learning from best practice in the West.

Lu Wei, China's former Minister for Cyberspace, admitted to having a hard job. Trying to control Chinese internet usage, he said, is like 'trying to nail Jell-O to a wall'. But China is not alone in making the attempt. He made sly note of the fact that Western countries engage in just as much interference as China, and added: 'The Chinese government learned its methods from developed western countries, and we still need to learn more.'[8]

Back in 1992, the China Council for International Cooperation on Environment and Development ('the Council' or CCICED, pronounced 'sea-said') was established by the State Council of the Chinese government. Its mission is to foster cooperation in the areas of environment and development between China and the international community. It consists of 25 Chinese members and 25 from the international community, 'chosen for their experience, expertise, and influence'.

'For China to develop well,' the Council's Chair, Zhang Gaoli, told the Council's AGM in 2014, 'we have to draw modestly on all the best practices around the world, and to learn from all the civilizations around the world. Only by doing that can we develop well.'[9]

Bands of Chinese officials went roving to find global leaders in public interest environmental law. In the summer of 2014, one panel of Chinese policy advisers reached Brussels. The group was headed by Wang Yuqing, the former Deputy Environment Minister of China, and included a member of the National People's Congress.

They were guided by a Dutchman, Dimitri de Boer, who ran the EU-China Environmental Governance Programme. 'They came in with specific environmental laws that were being drafted,' Dimitri recalled, 'and they told me some of the key sticking points and points of controversy that they wanted to have an international sounding board for. I brought in experts like James to answer those questions.'

James cut short his summer holiday and trundled by train from southern France. There was no clear reason to do so. Governments in Europe were largely antagonists in the environmental arena; the likelihood of China seeking genuine dialogue with a representative of a European NGO was slim. Still, the delegates had come a long way from

China, and James headed a Brussels law office. Some show of welcome was polite.

As James spoke to the assembled group, he saw some faces become alert. He had set up and run public interest law programmes in the USA, and battled to instil them into Europe. He spoke of bringing a hundred cases to enforce the Clean Water Act in US Federal Courts. 'They were intrigued,' James said. 'And then I explained how those cases had a deep, ulterior motive. By doing the government's work for them, singlehanded, I meant to embarrass them into resuming the work themselves. That's what happened. How the delegates loved that; they laughed and laughed. It perfectly fit the Chinese notion of "face", shaming the government into doing its duty by doing it for them. And then when they stopped laughing, their questions started. Real and good questions.'

They asked how NGOs are funded. Wang Yuqing explained how they were eager to get more NGOs going in China. Could such NGOs work if the government supported them or would they lose their independence? How would that situation work in the West? The government's study of how to kickstart a new generation of Chinese environmental NGOs was clearly in earnest.

The delegates returned to China. Shortly afterwards, an invitation from the EU-China Environmental Governance Programme arrived in James's London office. Would he fly to Beijing and share his experiences with judges of the Supreme Court?

This was for September 2014. China's revised Environmental Protection Law was about to take effect, and a judicial interpretation was urgently needed to give courts a framework for how to deal with cases brought by NGOs.

Amsterdam: sleazy sex for kids to gawp at in its red light district, and easy weed over the counter in its cannabis cafés. The city has a natural pull for teenagers. It wasn't working for Dimitri. He grew up there. Aged 15, he flew on vacation to Thailand, and a whole world opened up.

Decades into adulthood and there's still something about Dimitri that says 'Dutch' — the high forehead perhaps, the straight nose, some pert eagerness about his posture, like life is ever set to present his mind with something to which it should react. That's genetic patterning and cultural conditioning. However, when a 15 year old flew from Amsterdam and stepped out into the heady swirl of urban Bangkok, when he then moved out to the tropical balm of a health and meditation spa, his Northern European interior thawed.

Back at college, teachers were old school, preparing pupils to find their way in the world. Dimitri already knew his destiny; those weeks in Thailand taught him. He would become an Asia expert, and live in Asia. Education was simply getting in the way.

At 17, he quit and shipped himself straight back to Thailand.

Being young and in Thailand was enough for some years, though early on in his stay he did development work with villagers in the surrounding hills. A return to the Netherlands brought him a Masters degree and some corporate experience, but before long he went back to Asia, this time to China, where he studied Chinese and worked with the UN development agencies.

He was with the UN in Beijing in 2008 when the state environmental protection administration upgraded itself to become the Ministry of Environmental Protection. The new status prompted self-examination. What should the new Ministry's priorities be?

Dimitri's opinion? They should focus on transparency and public participation. It would help them do their job, strengthen their mandate, and strengthen what was a weak political position. Other experts concurred.

At the same time, the EU-China Environmental Governance Programme was formed, backed by EU funding. Its main thrust from the EU's side was the importation of Aarhus Convention virtues into China: access to information, public participation, and access to justice. Dimitri assumed leadership of the Programme in 2014.

He was interested in the project, if not overly hopeful. Back in Europe, the European Commission and member states deployed top lawyers against ClientEarth's, to spare themselves putting those Aarhus ideals into practice. So how ready was China to adopt such democratic fundamentals? At the outset, Chinese environmental governance seemed akin to a mountain clad in forest. You can entertain yourself for months while walking around it, and yet never find a path that leads up its sides.

Dimitri, in his new role, donned a suit and tie and entered his first meeting. It was by government invitation and stacked with government officials. You wear a wary smile for such occasions, and expect little. A woman from the Ministry of Environmental Protection rose to give the first presentation. Her topic was the need for better environmental governance.

'In terms of the level of knowledge, and how open she was about the need for this,' Dimitri recalled, 'she blew me away completely. She spelled out the problems so clearly. She said the government credibility has reached a new low. The country has thousands of mass incidents, people are in violent protests over environmental issues, the government is withholding information even though it's so obvious to everybody it should not be restricted, and they're just protecting the polluter.'

Her Powerpoint included cartoons snipped from Chinese media. In one, a government spokesman stands beside a huge old fashioned bomb flimsily veiled by a scrap of cloth. 'There's absolutely no problem here,' the spokesman announces to a rack of the media's microphones.

'I was looking at all this and thinking "What? Is this for real?"' Dimitri said. 'The next guy came up and he was from the local People's Congress. I imagined he was a rubber stamping apparatchik, and then he used the adjective "democratic" five times in his speech to describe the benefits of enhanced environmental governance. I didn't even know a guy like that was allowed to use those words. It was amazing. For me to see the level of understanding and the level of confidence, the obvious progress that was being made in this regard, I was just like, "Wow. Wow." I was really blown away.'

He paused with the memory, and then shook his head a little and smiled.

'And I'm still blown away by it. It's kind of a perfect storm where environmental governance could actually not only take very rapid progress but also drive further social progress in China. It really is now one of the most progressive government fields in China.'

During the Cultural Revolution, China felt no need of lawyers; from the 1970s, the legal profession entered a growth spurt after a standing start. Between 2004 and 2014, the number of certified lawyers rose from 30,000 to 270,000. That's fast, but it still means China fields just one lawyer for around every 5,000 citizens. (In England and Wales, the factor is around one lawyer per 300.) Since those lawyers like to base themselves in China's big eastern coastal cities, up to 10 per cent of inland counties have either only one lawyer or none at all.[10]

More and more lawyers are emerging from China's Schools of Law. They are still relatively few, and their very newness means they are inexperienced. From this starting point, they have been charged with supporting one of the most fundamental societal shifts in human history.

Humans have progressed through 'primitive civilization, agricultural civilization, and industrial civilization',[11] the latter now at the end of a 300 year run of unsustainability. China has decided to articulate and force into being the next era of humanity's planetary existence. It has pledged to bring into being 'ecological civilisation'.

The phrase was coined by the agricultural economist Qiang Ye in 1984, and became the substance of rigorous debate in Chinese academic circles.[12] That debate retrieved the concept from Ancient Chinese religious and philosophical discourse. For example, 2,500 years ago, Laozi wrote: 'Man takes his law from the Earth; the Earth takes its law from Heaven; Heaven takes its law from the Tao. The law of the Tao is its being what it is.'[13]

'Ecological civilisation' entered politics in a report to the 17th National Congress of the Chinese Communist Party in 2007. The then President Hu Jintao explained, 'building ecological civilisation, in essence, is building a resource efficient and environmentally friendly society that is based on resource and environment carrying capacity, complies with natural laws, and aims at sustainable development'.[14]

The phrase became enshrined in the Chinese Communist Party's Constitution in 2013, at the Third Plenary Session of the 18th Central Committee. A 12,000 character document followed in April 2015: *Central Document Number 12* fixed a programme of actions and targets. 'It systematically addresses the obstacles to effective policy,' according to the bilingual environmental website chinadialogue, 'by setting out standards, mechanisms and assessments that aim to improve implementation and realise the ambition it proclaims'. Officials would bear lifelong accountability for the impact of their decision making on the environment, while economic growth would no longer be 'the only criterion in government performance assessment'.[15] For the current leader, Xi Jinping, the forging of an ecological civilisation is a cause 'benefiting both contemporaries and future generations'.[16]

In a parallel shift of focus, in October 2014 the Communist Party pledged to establish the 'rule of law' by 2020. *The Economist* judged the Party's main aim to be 'to rein in local officials whose routine flouting of the law causes public anger and many thousands of protests every year'.[17]

The progress in bringing fresh environmental laws had already been huge and 'in many ways unprecedented in history'.[18]

China's foremost environmental lawyer Wang Canfa has remarked, 'In the past 30 years, China's modern environmental law has developed from nothing to an independent and important section of China's law system.'[19] Aligned to a university, Wang Cafa founded the Centre for Legal Assistance to Pollution Victims, 'the most influential environmental law NGO in the country',[20] which counsels and acts for citizens with pollution torts claims, and offers legal training. Its affiliate is Beijing's sole public interest law firm, Huangzhu Law Firm, which uses for profit cases to sponsor pro bono environmental work.

Environmentalists and environmental legal professionals from the West have helped 'guide and open the progress of China's environmental law'.[21] They can continue to do so — though without any experience in building an 'ecological civilisation', since that is revolutionarily new. Also, they must accept a categorical Communist Party on their plans for 'rule of law': 'We absolutely cannot indiscriminately copy foreign rule-of-law concepts and models,' it stated, with the Party seen as admiring the Singaporean model of 'benign authoritarianism'.[22]

That *Central Document Number 12* gave guidance as to the way forward. Xu Shaoshi, head of the National Development and Reform Commission, has noted: 'One of the breakthroughs of the new document is the idea that everyone needs to participate in constructing an ecological civilization … because the process needs the joint efforts of all.'[23]

When lawyers are too few, civil society must be empowered to take over much of the enforcement role. The task of ushering in an ecological civilisation is monumental. It needs all the new laws going, and every enforcement mechanism. Lawyers and judges need training, and citizens need fair and generous access to information and justice.

James flew to Beijing alongside two senior European judges. Their objective? To address a full day's meeting at Beijing's Grand Hyatt Hotel. James's particular task was to share his wealth of experience with public interest environmental litigation.

The meeting turned out to have a dual purpose. The morning session focused on judicial interpretations for the Supreme Court. In the afternoon, attention switched to members of the National People's Congress. They attended because they were finalising the law that allows prosecutors to bring public interest cases against local governments.

'It worked out really well,' Dimitri recalled. 'Senior representatives from both the Supreme People's Court and the National People's Congress Standing Committee met James, to discuss how they should set up the system for environmental public interest litigation and a trial for administrative public interest litigation. James was the star of the show. Two legislative documents were issued in the last months of 2014, incorporating the key recommendations James had made — particularly regarding setting up the rules for cost bearing in such a way that NGOs would only lose a little if they lost their case. It's hard enough to build a strong case, and it's unrealistic to expect that an NGO would be able to afford the legal costs of the defendant.'

The Beijing training day brought James a top level request — would he bring his work to China?

One sad fact of public interest law is that requests stop at that. There's seldom 'and here's the funding' or 'we'll make it worth your while'. James's ultimate destination: a private meeting in China's Supreme Court. First stop on the fight there: Hong Kong, and a round of fundraising and awareness raising meetings.

Beijing swaddles itself against the cold in sub-zero winters. In February, Hong Kong is on the springtime side of balmy. You have to work hard and climb a hill for a touch of greenery, but it's a pleasantly warm place to tune in.

After a series of advisory meetings and fundraising pushes, James and his then Chief Operating Officer, Cory Edelman, boarded a flight two and a half hours north to Hangzhou. Half an hour's taxi ride from the airport took them to the Alibaba campus.

Alibaba, which dominates China's vast ecommerce landscape, was fresh from completing the largest and most successful IPO in history. Its shares sold on the New York Stock Exchange for $21.8 billion. Senior staff made a mint. Cory Edelman knew them as friends from years back, when founder Jack Ma was moving out from being an English lecturer in the city and Cory was Vice President of Disney, looking after their Asia divisions. Today's meeting was with the head of the Alibaba Foundation.

The largest of Alibaba's three campuses, this one in Hangzhou borders the biggest wetlands park in China. Buildings rise through storeys of plate glass, fronted by atriums that frame the sky. Young executives speed through its streets on bikes. The place has the sense of an elite university, where everyone is young and smart and working hard.

The campus feels self-contained, with options from Starbucks to fine dining. Alibaba majors on digital commerce, but, for the old fashioned end of business where goods come through the mail, they have their own post office. It once received a car, parcel after parcel, some tiny and some bulky: nuts, bolts, carburettors, an engine, a window, a door, a tyre. Every part of a Lamborghini Diablo was tracked down online, using the network of Alibaba merchants, and an order was placed. Piece by piece, the car was assembled. Bright orange, the Lamborghini is sleek in a forecourt. It comes with a key and a key fob, all ordered online. Turn the key and the car is set to drive away.

Shen Bin, the Foundation Director, was in his 30s, a venerable age for an Alibaba employee. He led James and Cory straight to one of the smarter restaurants. 'We had this super lunch,' James recalled, 'and just cut to business. I made it clear, as I have learned to do in China, that we don't want to cause conflict, we want to solve problems, and one of the ways you do that is by cooperating with the government.

'They were very keen on training judges, particularly Supreme Court judges. They want to train journalists. When the journalists are reporting environmental stories as being about the health of the ecosystem and the health of the people, rather than as social unrest stories, then the judges will see the news, and see these are issues of a different type.'

The conversation shifted venue to a conference room. One of its walls was entirely made of glass backed by white: the whole wall was a whiteboard, splashed with ideas.

The lawyers and the funders buzzed around the idea of forming academic centres of excellence, to support training of environmental lawyers in the universities.

'NGOs are going to bring their own cases,' James added, and then went on a roll. 'Engineering expertise, biological and legal expertise, they will need all that. Let's create a consulting group of experts who can help in these environmental cases, whose services would be free. We could say they are working as friends of the court. They'd be there to support the courts in making excellent judgements.'

'Did you just come up with that now?' Shen Bin asked,

'During this meeting,' James admitted. 'You know, I looked at the wall, and I said, "This is a place for good ideas."'

'You must come back and have more of these good ideas. That's a great one. We want to keep thinking with you.'

The Soviet Embassy used to stand just off Tiananmen Square. With the end of the entente between China and the Soviet Union, the Supreme People's Court moved into the building. The old Embassy was then demolished and a new courtroom rose in its place.

'You know, the US Supreme Court is really quite small,' James says of the comparison. This Chinese version has really affected him. His best way of explaining it is to repeat himself. 'It's an incredibly beautiful, modern, huge building. A very beautiful, vast, enormous building.'

Its atrium is five storeys high, dominated by a single sculpture: 'A picture of mountains; a bas relief, made with granite from all over China, beautifully done. The whole thing is a metaphor, so there are also waterfalls in the depiction. The sky is gold leaf up top, and a phrase on it runs roughly, "We look up to justice like we look up to a mountain." And, "Mountains are firm like granite, and so is justice; and justice also has the softness of waterfalls." The poetry of that was lovely.'

The meeting room was equally impressive; a chandelier cast light off a white and brown marble table some 4.5 metres wide and 12 metres long, which was surrounded by white leather chairs. James and Dimitri de Boer of the EU-China Environmental Governance Programme sat with a translator on one side, and a grouping of Supreme People's Court judges and advisers on the other.

Chief among these was Wang Xuguang. He can be utterly charming, with a Buddha-like sense of beneficence and calm, but you don't mess with Judge Wang.

Judges start young in China. It's a career choice, and Wang Xuguang made the step aged 20 in 1985, when he graduated from Shandong University. Compared to top flight lawyers, judges in China are poorly paid. Esteem was pretty low too, since to some extent they were seen as government functionaries. It takes a fierce mind and spirit to help forge an independent judiciary. By 2013, Judge Wang had become Deputy Head of the Jinan Intermediate People's Court. The high profile trial of Bo Xilai was moved into his jurisdiction.

Bo was the Communist Party secretary in Chongqing and a member of the 25 person ruling Politburo, the highest ranking official to be brought before the courts in three decades. 'Sina Weibo, a leading Chinese social network, published the trial transcripts. In the weeklong trial, the court account attracted 582,000 followers. Its 160 posts brought tens of thousands of comments. The added transparency helped render Bo, at least in his official statement, 'more confident of the future of China's judicial system'.[24]

'What Bo did has caused great losses to the interests of the country and people, and the circumstances were especially serious,' Judge Wang declared in summary, and handed Bo a verdict of life imprisonment — for corruption, the theft of public money, and the abuse of power — and the forfeiture of all his private assets. Mo Yuchuan, a law professor at Renmin University of China, studied the case and read the signals. 'The verdict came as the Party is gearing up for the upcoming Central Committee meeting in November,' he said. 'It is a sign that the leadership's unwavering determination to fight corruption will be continued.'[25]

Judge Wang was promoted to the Supreme People's Court in 2014, and was soon placed as the Deputy Chief Judge of the Environment and Resources Tribunal with that top court.[26] That sent its own message, reinforced with his 2016 elevation to the Chief Judge role. The judge who had dispensed justice against a top official was set to be as resolute on behalf of the environment. James and he had met during the Supreme Court training the previous September. Wang now led the judges at the meeting. James enjoyed the true warmth of his welcome.

'We talked for an hour,' James said, remembering the encounter. 'We talked about exactly what we wanted to do. Judge Wang said he was so happy to have met me in September, and it was great to have this meeting because we would get to know each other better.

'He said he was also very pleased that I had followed up on that meeting in September by giving such excellent comments on the Supreme Court's written regulations, which was called an interpretation, of the citizen suit provision of the new environmental protection law. He said my comments had helped their drafting, particularly on the cost provisions and the burden of proof.'

Those comments included extensive case studies sent after the meeting, which included records and analysis of James's work on the Citizens' Enforcement Project in the USA.

Early in his UK career, for one tremblingly exciting moment, James thought the British courts might apply a similar cost shifting rule to their jurisdiction as was applied in the US. The senior judge charged with reviewing the process, Lord Justice Jackson, came to that conclusion in an early draft. But the UK blanched before so progressive a move. China, in a momentous move to boost civil society's abilities to enforce environmental laws, encoded such cost progressive rules in the implementation mechanisms of the revised Environment Protection Law.

'James made a very strong case about one way cost shifting to the Supreme Court,' Dimitri remembers of that September training day. 'I think it had an effect. We can look at feedback from them. Some of the Supreme Court people were very candid about telling James which areas they followed up on and built into their system, and they actually did say such things, they were very grateful to James for a bunch of things he said.

'What I see most clearly in the Supreme Court's judicial interpretations, its clarification of how public interest litigation should work, is that they are very conscious of the difficulties that NGOs face in bringing such cases — particularly financial but also in terms of their lack of capacity. They have interesting clauses that clearly favour and protect the NGO from excessive costs and allow them to get cost recovery.

'Also, if they come up with a case that seems compelling but for which they don't have very adequate evidence, the courts are in a position to collect that evidence. So the courts can conduct their own investigations. Essentially, they are taking over part of the technical function that the NGO is perhaps unable to fulfil.

'It's funny for the courts to become so engaged in a case and essentially take sides, but it's irrefutable. In some areas, judges will go out and physically supervise the evidence collection. If the court takes the evidence, who are you to argue that the evidence is wrong?'

Inside the Supreme Court in February 2015, just a month after the revised law came into force, Judge Wang gave whole hearted approval of ClientEarth's plan: among other work, they would continue the judge training that the EU-China Environmental Governance Programme provided alongside the Ministry of Environmental Protection.

'James, just please learn from Dimitri,' the judge said, 'he's been doing it so well. Carry on the way it's been doing, and do more of it, and we'll be extremely grateful because it's so valuable. The Supreme Court are getting trainings, but all judges need these trainings, so please do that with us.'

Justice Zheng Xuelin, at that time the Chief Judge of the Environment and Resources Tribunal of the Supreme People's Court, gave a title to his address on the day China's revised Environmental Protection Law came into force: 'Spending Ten Years Polishing a Sword and Showing It Today.' In the address, he expressed the hope that this 'powerful' sword 'can cut through the dirty stream and clean the grey smog air. It will be like a sword of Damocles that hangs above the polluters.'[27]

Particularly striking is the 'ten years' of polishing such a sword. That change to the cost rules was the final sharpening of the blade. China prepared for battle on behalf of the environment and public health for more than a decade. Actions now are part of a strategy assembled in corridors of power. It is those years of planning that have built the current momentum. The scale of the challenge is vast, and the speed of the manner in which it is being addressed can be dizzying.

A Western organisation going to China needs to adapt to a new way of working. When in Beijing, do as the Chinese do. ClientEarth's Warsaw office was established as a subsidiary organisation with its own Board, and staffed by Polish lawyers. Any opposition to the government's environmental actions was subsequently taken from inside the jurisdiction. In China, ClientEarth's lawyers would serve as foreign experts invited to support government action.

James stated it simply in his Supreme Court meeting with Judge Wang. 'The government has already taken all of these remarkable steps for the environment, and therefore I want to help the government implement these.'

'You will be extremely warmly received by the government if you want to do these things,' the judge replied. 'Very, very warmly received.'

Dimitri turns to an old Chinese expression to explain how one approaches working alongside the Chinese government: *If you're standing straight, you don't need to worry about your shadow being crooked.* 'It doesn't matter how it looks. If you know that you're fine and right and clean, you don't need to fear. We're working openly with certain government departments, which gives us some level of shielding from those whose interests are threatened by environmental law enforcement. Where you are pioneering something, it's often cleverer to start with small steps. These steps might seem small to some observers, but they are actually very big in the larger picture of things.'

One such step happened in May 2015, with the new Environmental Protection Law still in its first flush. Friends for Nature, the foremost Chinese environmental NGO based in Beijing, teamed with the local NGO Fujian Green Home to file a complaint against three individuals already imprisoned for illegal mining. Previous torts litigation looked for personal damages, but this broke new ground in seeking reparation for damage to the ecosphere. Experts visited the destroyed site to assess the range of its tree species, the damage to the vegetation and soil, and its value as wildlife habitat. Costs were assessed for restoration of the land, and for what the years of damage caused in loss of 'ecological function'.[28]

The case set precedents, including the standing of NGOs to bring such cases and the award of attorneys' fees. Transcripts of the trial once again featured on the Sina Weibo network, as part of building public confidence in such trials actually making a difference.

In 2014, the first year of the revised Environmental Protection Law, the courts accepted 48 environmental public interest cases. They came from 16 plaintiffs, including a number which were backed by local People's Procuratorates. (The People's Procuratorate, 'China's counterpart to the U.S. Department of Justice and state prosecutors' and hitherto largely 'uninvolved in pollution enforcement',[29] is another of those groups on the docket for training support by ClientEarth's China team.) The China Biodiversity Conservation and Green Development Foundation (CBCGDF) led the field among NGOs, with 14 cases proceeding. It was a start.

'For most NGOs bringing a case like this is like eating a crab,' says Yanmei Lin, Associate Professor at Vermont School of Law, 'you just don't know where to start.'[30] NGOs have to have been registered in the field for five years to have standing. Only a handful have experienced environmental lawyers on staff. There's a temptation to wait and see how well others do.

Public interest law groups in the West have effective models of working practice to share. Another way of moving forward is to understand what blocks the transport of successful rule of law from one jurisdiction to another. In the USA, for example, ownership of natural resources such as rivers and forests is either public (owned by the government) or private; the owner has the right to bring litigation. 'It's different in China,' Yanmei Lin observes. 'Even within one nature reserve the rivers and forests belong to different authorities, while ownership of groundwater and mineral resources is also divided.'[31]

These are the same issues that face ClientEarth's West African lawyers; they form a bridgehead between civil society and government, ensuring full stakeholder participation in the clarification of natural resources ownership.

Environmental law has become an unusual democratic force in China. New developments now require environmental impact assessments, and these require a level of public participation. People find it hard to believe the request to share their opinions is genuine. Nobody has asked them what they think before, and who would be interested anyway? Then they start talking.

'Typically, they know very little about environmental impacts,' Dimitri noted with some delight. 'They'll complain about all kinds of economic or social impacts. Ninety per cent of the comments that come in through the EIA [Environmental Impact Assessment] process are non-environmental issues, because this is the only time you get to ask this villager what she thinks about anything. It shows how pioneering this level of public engagement in environmental protection really is.

'You see that in access to justice too, NGOs being allowed to bring cases for the public interest in environmental protection. That's a pretty big deal. I don't think many other laws and regulations specifically raise NGOs as a suitable litigant. Traditionally, they prefer not to encourage civil society development, but here an open and very meaningful role for civil society has suddenly been created.'

China is openly encouraging the extension of citizens' rights to environmental justice. 'Coming to the human rights issue, China attaches great importance to human rights,' Xi Jinping told a UK news conference in October 2015. 'We have found a path suited to China's conditions. There is always room for improvement in the world.'[32]

In the same month, the Federation of European Bar Associations 'expressed grave concerns over lawyers that have been detained and threatened by the Chinese state. In July, approximately 225 lawyers, staff, human rights defenders, and their family members were arrested in China. Approximately 30 are either still detained or missing.'[33] In

January 2016, lawyers and jurists from Europe, Pakistan, Australia, and the USA signed an open letter to express their 'deep concern about the scores of lawyers detained or intimidated in China'.[34]

This makes for a heated climate in which to support the growth of public interest environmental law in China. It requires a nuanced reading of what might be achieved, and a judgement that any effort might leverage significant impact. 'We're being asked to help shape the laws,' Dimitri de Boer explained. 'In order to be invited to that party, you can't be too controversial. If things move too fast, it can cause unnecessary delays.'

As evidence of the success of that approach, Dimitri noted the Supreme Court's willingness to sign a formal cooperation agenda with ClientEarth. 'At a time when overseas NGOs are challenged, the Supreme Court is inviting a gutsy international NGO right into their group of inner advisers. I think that's a breakthrough.'

Voices that speak on behalf of the environment have to insinuate their arguments among separate vested interests. This happens everywhere. I interviewed Elena Visnar Malinovska, for example, as she completed five years in the EU Cabinet of the Commissioner for the Environment and was set to move on. She saw advantages in members of her team taking their expertise into different posts and working to 'green' what they were doing.

'I wouldn't be against going, for instance, to the Directorate-General Enterprise and Industry,' she said, 'to look from their shoes at how they see it, and whether some of the arguments [against strengthened environmental governance] might be well founded, or are just the usual advocacy for the incumbents.'

Zhang Gaoli twins his role as Chair of CCICED with that as Vice Premier of China's State Council. Environmental ambitions are filtered through the continued drive to lift the Chinese population out of poverty. 'Everyone should understand,' he reminded the CCICED members, 'we are still a developing country of 1.3 billion people. Without development we cannot create jobs for everybody.' He envisaged 'a benign circle of development. Otherwise GDP goes up at the expense of the environment. It's not what we want in terms of development ... Everything has to be transparent. Transparency, public participation, and supervision are important.'

He also spoke of coal, the vested interest that squats on so many ecological ambitions. 'We plan that by 2030 non-fossil energy will count for about 20% of primary energy consumption,' he said. That was the bright side. And then there was the dark side. '[US Secretary of State] Kerry asked me to offer more, but we know that China's energy mix is coal dominated. 67% of our energy consumption is coal, and coal is a big part of our resource. Given our resource endowment and energy mix, we can only make a commitment that we believe we can deliver.'

Continued dependence on coal is clearly anathema to the establishment of an 'ecological civilisation'. China has to achieve the bruising acceptance as faced by Poland and other nations: that coal is a resource that needs to stay in the ground.

In the meantime, Zhang Gaoli focused his talk on the need 'to speed up the construction of the legal system for ecological civilization ... We should pass new laws, revise or abolish existing laws, and put in place a sound legal system regarding the eco-environment, land, forestry, and pastures.'[35]

The phrase delineates one of the most vital tasks for the new millennium: the construction of the legal system for ecological civilisation.

'China is one of few countries in the world to have a long history in drafting environmental laws and regulations,' writes Bie Tao, a senior official of the Policy and Law Department of the Ministry of Environmental Protection.[36] That history stretches back through three millennia. Bie Tao cites a mention in the Han Feizi, written 475–221 BC, of a law that held sway under the Shang Dynasty, 1600–1046 BC. This required a 'cutoff' for anyone found leaving rubbish on the highway

— a 'cutoff' requiring the forfeit of the perpetrator's finger.

In 2015, Luo Guangqian, the head of China's first environmental court, was alert to new challenges. 'The ideal environmental judge has a background in science and engineering, allowing them to more easily understand the technical details involved,' he said. 'But such candidates are scarce.'[37]

Bounties are still to be discovered in China's new environmental laws. A new import to China is the use of expert testimony as evidence. 'An expert on soil pollution might now be called to testify on whether pollution can be cleaned up, or how much damage has been done,' Luo Guangqian said. 'A court could then accept that evidence, rather than commissioning a third party body to carry out an independent assessment, saving both time and money.'

The Children's Investment Fund Foundation, based in London, offered a major grant to fund the extension of ClientEarth's work in China. Dimitri de Boer transitioned out of the EU-China Environmental Governance Programme to lead ClientEarth's efforts in China.

Dimitri has a Chinese wife and two young daughters, who can already chatter away in Mandarin, Dutch, English, and French. The family is international, and when Dimitri looks out of his Beijing window on days of smog, his concerns are distinctly local. This is the smog that will invade his daughters' lungs. 'I get emotional regularly,' he admitted, 'because the stuff we're talking about is life or death for many people and for much more than people. I think it's worth emotionally investing in something like that.'

The US environmental groups NRDC and EDF have run Chinese operations for some years. 'Environmental law is not their core business, it's part of their portfolio,' is how Dimitri distinguishes his new role. 'ClientEarth is dedicated to the core business of developing the rule of law in China. It's unparalleled.'

Judge Wang asked James to continue the judges' training. In June 2016, the Supreme People's Court called 300 environmental judges to Beijing for their annual training. James flew in and added the sum of his experience, joining as a lecturer, and bringing in other Western experts.

Earlier that week, James co-led a seminar on climate change litigation in the Supreme People's Court for its members. It was the first time that foreign experts were invited to a seminar in the Court's building.

After that seminar, when speaking to the 300 assembled judges, Judge Wang made a point of saying that climate change litigation, broadly defined, should become a priority for the Chinese judiciary.

Public interest environmental law groups began in the USA around 1970. James injected the model into Europe, and ClientEarth lawyers developed the techniques of promoting civil society's use of law in Poland and then in Africa. The extension of this work to China marked a remarkable waypoint in the movement's global journey.

One year on from the introduction of the Environmental Protection Law, Judge Wang's summary of what the judges needed to accomplish was particularly keen. 'Judges must change the way they handle environmental disputes, modernize their skill sets and stop favoring businesses over damaged ecology,' he told the English language newspaper *China Daily*. 'With environmental disputes increasing rapidly, the need for judges to adapt to revised laws, new case precedents and to understand their role in environmental protection has become more acute.'

That role requires new skillsets. 'Damage evaluation is a big challenge in tackling environmental disputes, as it is a technical process that involves complicated calculations,' Wang said. 'Assessing the damage and how much money is needed to remedy a situation requires a great deal of time for judges to reach a verdict.'

That time consuming evaluation helped explain why few verdicts had been reached in environmental public interest litigation. A principal evaluator of the performance of Chinese judges is the number of cases they conclude. Since environmental cases come with so many added complexities, their performance needs to be reviewed in new ways. Training would clearly help.

Judge Wang's fearsome practicality, and China's determination to swing environmental civilisation into being through the rule of law, is neatly captured in the judge's concluding statement. 'Difficult damage assessments are not a fundamental obstacle to dealing with environmental disputes,' he said. 'After all, what we do is to judge. When the damage is difficult to account for and the evaluation cost is too great, we must still give a verdict.'[38]

Doing deals for the Earth

James Thornton

People in Europe had a hard time understanding what I wanted to do. For some, it made sense immediately. For others, though, even detailed explanation failed to convey the vision. This problem was new to me. In California and New York, I'd found instant understanding.

I worked to make the message clearer. That is always a good thing. There were still problems. I realised that lawyers had not done this kind of work in a sustained way in Europe before, so there might be a cultural perception they should not. Lawyers in much of Europe are the people who do your property transactions and your wills. It was hard to see them as world changers.

The breakthrough came in talking to my trustee Howard Covington. He is a successful financier, with experience of doing corporate deals on both sides of the Atlantic. He said, 'James, it is very simple. Here is how the difference in culture works. In London, bankers do the deal and bring lawyers in to help. In New York, lawyers do the deal and bring bankers in to help.'

Shortly after, I was having dinner with a group of former McKinsey partners. I talked about my work. Despite their sophistication, they weren't quite getting it. So I told them Howard's idea, that we were the kind of lawyers who did deals — in our case, deals for the Earth. Light bulbs went off in the minds of everyone around the table. They started asking how they could help.

That image stays with me. We are deal making lawyers for the Earth. When you do a deal, you understand your target, figure your route, and understand the rules. You bring together all the parties and all the elements you need to make the deal happen. When it works, there is a monetary reward.

For us, the reward is protecting the living things of Earth, but the rest is parallel. We define our goal, survey the route, and bring together all the elements needed to reach it. Depending on the goal and the arena we are working in, the elements will differ. Here are some examples.

In Poland, our aim was to help move the economy to sustainable and secure energy. We realised that the proposed building of 14 huge coal fired power stations would replicate the Soviet energy policy of the 1970s and lock in high carbon for 50 years. So we challenged these investments and worked with the media to get out the message that there was a better way. We worked with local people who needed help in opposing coal plants. A milestone was passed when Energa, one of the largest utilities, announced that it would instead build a gas plant and improve the efficiency of the grid, reducing the need for new coal.

In finance, the question is how to move capital to fund the carbon neutral economy. The Paris Agreement set the contours. Now the system needs to deliver some $90 trillion over the next couple of decades to build a decarbonised economy. We are working with investors, pension funds, economists, progressive companies, and other stakeholders to envision how the system needs to shift. We are engaged in foundational work to quantify the risk of climate change to assets. This will help trigger fiduciary duties. We are looking at models of corporate governance that would allow companies to make the right long term investments. Rapid change in the right direction is possible, and we are working on it.

In Brussels, we saw the need to improve the Common Fisheries Policy, then under revision. Scientists said that unless the law was improved, fish stocks would crash. We brought together scientists and lawyers, and designed policy that could help. We worked with NGOs and lawmakers across the EU. We used media. We wrote memoranda for parliamentarians. We wrote amendments to the law. Before a crucial vote in the European Parliament, we brought scientists in to speak to lawmakers. In the end, the law is much improved.

In Africa, we want to help forest dependent communities, and specifically help them protect forests. The best way is to build legal capacity among the African actors. We work with lawyers and NGOs to identify how the legal systems treat the rights of the people, the rights of land ownership, and the rights to participate in decision making. In this way, we help build intellectual property in the community, so that the African lawyers and NGOs can be more effective in solving the problems they face. And it is working.

In China, the government is focused on solving the country's huge environmental problems. Here, the emphasis on systemic change is at a vast scale. The laws need improving, and enforcing. We are helping. We are training judges and prosecutors to help them bring good cases and decide them well. We will be helping NGOs build their capacity to bring cases against polluting companies, as their law now allows. When the needed adjustments are made, there can be great gains in dealing with pollution. This is the first step in realising their vision of building an ecological civilisation.

These are examples of where we act as deal making lawyers for the planet. The goal is always systemic change to protect civilisation and biodiversity. The means and the elements shift in the different arenas. Bringing the right elements together to do the deal, then closing the deal, is always the modus operandi.

10

The Judgement of Paris

November 2015, and terrorists set loose attacks in Paris. Suicide bombers self-detonated among football crowds at the Stade de France, gunmen sprayed bullets into customers at café tables, and scores of music fans were killed in a concert hall. Some weeks later, James walked down empty streets alongside the Tuileries Garden.

Delegates had flown into Paris from the tropics, from every state in Africa, from island nations awash with rising seas, from Arctic tundra, from Washington, DC, from Beijing. Bombs and rapid fire had turned Paris into a war zone. Now all the world's governments focused on the city, to work at healing the planet.

This was the 21st international Conference of the Parties, Paris COP 21. World leaders had graced the opening week, and it had a week still to run. The ambition was huge: agree measures that would halt the steady rise in global temperatures.

In times past, the corridors of Versailles provided achingly severe training in diplomatic skills. Diplomats worked in a global arena as France maintained and then released its Empire. Now a team of French diplomats headed by Laurent Fabius marshalled thousands of delegates towards an agreement when the Copenhagen talks of 2009 had failed to do so.

James had just laid out his aims for Paris in an open letter to *The Times*, co-signed by eight QCs and a range of legal experts and supporters.

Whatever happens at Paris, he urged, state leaders must return home and write their pledges into their national laws. Pledges must become legally binding, with an open route for citizen enforcement. 'It will allow citizens in all of your nations to hold you to your promise, and keep the earth more safe than before.'[1]

A courtyard shone in soft amber light. James walked past the Élysée Palace, home of the French President. Louis XV bequeathed the area its grandeur, with a carelessness that ushered in the French Revolution. James walked past the palatial British Embassy, built in the 1720s and once home to Napoleon's sister. Napoleon himself used to cut in through the gardens for assignations with his sister's lady in waiting. Now, black limousines passed and pulled up outside the building ahead. That was good. James would have an audience. This was his destination, the Cercle de l'Union Interalliée. One in a row of great 18th century mansions, it reconstituted as a social and dining club in 1917: American soldiers had marched in to liberate Paris, and this place was refashioned to make these allies welcome. It held a history of achieving peace through diplomatic relations.

There was a good crowd: senior bankers, and the heads of pension funds and environmental foundations. They drank cocktails beneath chandeliers while the last of their number arrived, then filtered through to the dining room. Ben Caldecott of the Sustainable Finance Programme at the Smith School, University of Oxford, the co-host of the dinner, chimed a glass and stood to introduce the speaker.

James watched an array of French cuisine land in front of every guest. His job was to speak. His own food grew cold. 'It was one of those moments when you knew you were speaking to a group of people as intelligent as any you will ever meet,' he recalled. With a wink of understanding, a waiter took his plate away.

I can't supply the meal, but I can give you a taster of the argument that ran across it, in four courses.

You expect an environmental lawyer to speak about environmental law. In fact, the legal ingredients of these courses are everything but. Company law, Corporate Law, State Aid Law, Fiduciary Duties, Torts, Competition Law, Energy Market Law — in sum, the issue is finance.

Include yourself among the group around the long table. Uniformed waiters ease fine wine into crystal glasses. You note the jewels in earlobes, the flash of a high end Swiss watch, the folds of silk. Your fellow diners are cusped in a living remnant of France's imperial era. They administer great wealth. Their attention quickens as James speaks on. Yes, they care for the planet, but they don't need more details of noxious air or a species on its knees. Saving the planet is a shared goal, but no one saves the planet without the funds to do so. That's what they provide. They deal in money.

And that's what they hear James discussing. Financiers play by a set of rules that would have been recognisable to the Medicis. Those rules are vested in growth and plunder, and are heedless of the sustainability of natural resources. The consequences have ramped up and are set to be unleashed. This century will be rocked by the effects of climate change.

That brings a whole new scale of risk to an investment portfolio.

It's like you are wrapped in furs, skating as a happy party across a frozen lake, and a voice faint with distance calls out from the land: 'The ice is thin and melting.'

Attention around the dinner table is rapt.

In China, a woman from the Ministry of Environmental Protection headed out into the provinces to train environmental protection officials. 'Look,' she began, and told them about the case of an official like them. 'This guy got put in jail. Why?' She led them through the stages of his fall. Very likely the mayor or the party official of that local municipality told him to OK the development of a polluting project. When everything went wrong, his was the head to roll. What's the minimum he might have done to ensure responsibility was placed at a higher level? The officials paid her profound attention.

These dinner guests are at that higher level. Some of them have their own point of acute vulnerability. They are fiduciaries.

And so to the first course. Join the board of a pension fund and you become a trustee. 'Trust' is the core of that word. You assume the trust of beneficiaries, the members of the fund, and look after their interests. The law of fiduciary duty sets out what you must do.

Consider the manager of a trust pension fund. One basic rule for the manager is to take risk into account. How far into the future should she look? Twenty two year olds pay into the fund and so do 60 year olds, and it must treat them equally. What's the world going to be like when today's 22 year old draws down her pension in the 2070s?

'No such risk is yet acknowledged in law,' James admits. 'We're going to change that.'[2]

Since pension funds are among the largest owners of equity, that change alone could redirect trillions of dollars' worth of value. The argument for change starts with fundamentals. Currently, global shareholdings amount to around $70 trillion. A very conservative projection of the potential damage from climate change sees a risk of around $7 trillion. So 10 per cent of the value of all asset classes are at risk. This is double the amount needed to qualify as a risk that the pension fund manager needs to take into account.

'Don't worry,' James says, mimicking a typical investor's response to the challenge. 'I am a smart person. When climate change is a real risk that's going to hit, I'm just going to switch out of oil into consumer products, or pharmaceuticals, or something else.'

He flips back to speaking as himself. 'That is the typical response: there is no risk that I can't compensate for by the normal methodology I use as an investor. That's the argument we have to kill if we're going to win the fiduciary duty argument, and require that all major investors have to take the climate change risk into account.'

The argument James and his colleagues are building is that this risk is structural: because of the nature of climate change, all classes of financial assets will be affected. There will be no place to hide.[3]

A provision in UK law allows a pension fund trustee to approach a court and say, 'I believe I may have a duty, but I'm not sure. Can you instruct me?' The search is on to find a pension fund willing to partner with ClientEarth's lawyers in such an action, and establish climate change as a risk the trustees must factor into their investment strategies.

The search is hard. 'More difficult than I thought,' says James, and casts his eye around the pension fund representatives around the dinner table. 'There's a lot of natural history involved, observations of my fellow mammals. I assumed that because these pension fund trustees are very smart people, they would get this argument, and it would make a lot of sense. Some of them agree with it intellectually, but the idea of actually going forward and asking for it is difficult. Core duties of being a trustee are caution and prudence. We need that rare animal, the prudent trustee with a comfort zone as wide as an ocean.'

Course two: a neat little number called the Capacity Market Mechanism. James explains how the EU requires all member states to have a plan, approved in Brussels, to ensure constant electricity supply. Countries can direct state aid to companies that provide the so called base load. This is an opportunity for enlightened governments to support clean energy and for renewables to become a major contributor to an integrated system.

Britain was the first country to put its mechanism through. But away from the public eye, it did a backroom deal that favoured the coal industry. ClientEarth could not get into the EU courts, because of how the EU closes its courts to citizens. So it teamed up with a start-up company. This company wanted to sell ways of reducing demand for electricity, a cleaner way to keep the lights on. Because it could claim economic harm from the decision to use the government money to fund coal, it would have standing. ClientEarth also alerted the EU's Director

General for Competition that the UK mechanism violated competition rules. Britain duly announced it would reconsider its Capacity Market Mechanism.

'This is a small citizen effort that will lead to a more transparent, open process, in which the use of the Capacity Market Mechanism will be debated. Why is it important? It's a lot of state money that gets funnelled in one direction or another.'

Course three: a quick taste of internal energy markets. Hydro energy powers Norway, sun shines into Spain's solar fields, and winds turn Germany's turbines, but Europe is broken up into what are essentially energy islands. Monopolistic energy companies sit astride the grid, not wanting to change. The greatest of these octopuses is French and called EDF (Électricité de France). It is still vertically integrated: it owns or controls the method of delivery for its product. That is long since illegal in the EU, but the laws in place are flouted. It is also majority owned by the French government. So the government will do nothing, and the EU turns a blind eye.

James shares a dream, verging on an ambition, to bring an antitrust case against EDF to break it up. Starting the battle at a smaller scale, he sees lawyers supporting a French solar energy company in a competition law case. EDF owns the grid, so the company would need their permission to connect. Such a start-up has modest finances, enough for perhaps six months of operations. Months go by, letters go unheeded, permission is not refused but nor is it given. 'General Kutuzov used this strategy when he fell back and back and back against Napoleon, and the Russian winter ate Napoleon's army. The French seem to have learnt their lesson,' James says.

The grand Parisian setting suits the comment. 'What I would love to do is bring a competition law suit on behalf of such a company, force connections, and get a big anti-monopoly penalty. You would chip away at the monopoly. You cut off the dragon's toe rather that slay the dragon all at once.'

The fourth course: a slice of tort for dessert.

'Everyone loves tort,' James breezes as his introduction. 'For those of you who don't know yet, you will love it. It's the law of damages. I'm fond of a comment from the American writer Gore Vidal. An interviewer asked what was important to him. "When I was a young man, sex was really, really important," he responded. "Now that I'm an old man, I prefer a good tort.'"

The diners gust laughter. Good. It clears their minds for the true blast of a tort that is heading their way.

A tort case is an action for damages against somebody who has injured you. It's the classic argument among human beings from physical to reputational harm. How would that be useful in the climate context? James gives the example of one of ClientEarth's trustees who ran a €18 billion a year company. When they first met, the man told James: 'You need to bring damage actions against big companies so that the CEOs of these enterprises lie awake at three o'clock in the morning wondering about their personal reputation, about their personal liability, and the way that plaintiffs are going to destroy them and their company with damages.'

A target might be a company such as Exxon, which is credited with having produced 3.22 per cent of all greenhouse gas emissions. Could they be held accountable for 3.22 per cent of a personal injury from climate change?[4]

The main hurdle is attributing causation: say, for example, as claimant you find a woman whose mother died in the Paris heatwave of 2003. How might you prove fossil fuel emissions were the cause?

Science already gives a fair certainty to the link between carbon emissions and climate change. A new branch of science has emerged to prove the next stage: that a particular heatwave is caused by climate change. Climate Attribution Theory seeks to draw robust links between human contributions to greenhouse gases and individual climate events.

The likelihood of any big climatic event rates as a 1 in 1,000 year event, for example, or a 1 in 100 year event. James approached a lead scientist in the field and asked for robust data that showed an event going from a 1 in 1,000 year event to a 1 in 10 year. Would it then be obvious to a judge that the event was caused by climate change? The judge has to decide that an event was caused 'more probably than not' by the thing complained of. The judge needs to find that the balance of the evidence, meaning 51 per cent or more, shows causation.

'Get the right data, the right event, the right plaintiff, the right defendant, and the right judge, and you might be able to get one of these cases to work, which I very much hope to do,' James tells his diners.[5] 'The scientist re-analysed the data in a paper he had published a few weeks earlier. "I can do better than that," he told me. "We can now say the heatwaves in the South of France have gone from being a 1 in 1,000 year event to what we think is probably a 1 in 5 year event."'[6]

James finished speaking, His reheated meal was returned to him, his glass was filled, and he snatched food and sipped wine as folk asked him questions.

James took the lawyer Alice Garton with him, to answer questions on her specialty areas. 'There was a moderated discussion,' she recalled. 'People were so engaged with the topic. James had answers for everything, and brought people round who obviously didn't like the idea of litigation. It must have gone on for two or three hours.'

Alice was new to ClientEarth. This whole approach of being a public advocate for pioneering legal work was a novelty. Many of those she most respected in the movement were in the room. This event in Paris was at the heart of something. But could the whole dire momentum of climate change truly be halted?

A child places herself in the wider world when she learns to write her address. Here's what Alice learned to copy out:

Alice Garton
Solar Village
Humpty Doo
Northern Territory
Australia

Home was an earth house with no locks. Up in its ceiling, the children watched olive pythons and gave them names: Maisie, Slippery Sam, Lionel Long. Lionel Long grew ever longer. At five metres, they called in the snake catcher to take him away.

Pet wallabies bounced around outside. So did the children, because indoors brought the temptations of electricity. Since this was Australia's first solar village, set up by their parents, the children knew electricity was expensive and knew to save it whenever they could.

Alice spent hours out in the fields, picking beans for biofuel. She watched a wind tower rise and its blades stick. This was the 1980s. Those experiments in alternative energy were pioneering. As with many such experiments, they often failed. They were all part of Alice's schooling.

The teenage Alice had to take buses north to Darwin, the nearest city. First up was high school and then university. Alice needed a mobile phone. The saleswoman needed an address for the contract. 'Solar Village,' Alice began, 'Humpty Doo, Northern Territory.'

Alice was dismissed as a prankster.

Shift on a few years. Alice took a degree in Law and Politics. She reckoned that fit her for the likes of Greenpeace.

'Mum and Dad said no,' Alice recalled. 'They said to go and work for the best law firm in the commercial sector for ten years or so, and then make the move.'

That, they thought, would power their daughter with a genuine voice.

After four years of commercial litigation in Darwin, Alice left for London. She landed an internship at Chatham House, the top think tank, and worked on biodiversity, biotech, and climate change issues. That year of 2006 included David King's presentation on 'The Science of Climate Change', and *The Stern Review on the Economics of Climate Change*. It was heady stuff.

And then back to a proper job. As an in-house lawyer at Brookfield, Alice set up a climate change practice, advising boards of directors on the implications of the Climate Change Act and Energy Act, and looking for new business opportunities. 'That gave me direct access to the board, looking at the green retrofit market. We set up Brookfield Green, and did the first full energy retrofit of a building down on Canary Wharf — where BP and McGraw-Hill are, funnily enough — based on the retrofit of the Empire State Building.'

A break to give birth to twin boys was a chance to study part time at University College London for a Masters degree in Environmental Law. Finally, a job came up in an NGO. She was ready.

Follow Solar Village, Humpty Doo, onwards. As a biography, it's pretty unique, but then so is a fingerprint. Follow Alice into her new workplace in a public interest law group and you'll note that what makes her stand out also makes her belong.

ClientEarth's HQ is The Hothouse, stretched along one side of the parkland of London Fields. This is in Hackney in London's East End, once known for poverty, but now urban chic. Architects won awards for tear shaped windows that drop light on the building's workspaces, and the place retains a funky style. Parakeets raid feeders stuck to the outside glass, while the leaves of tall pot plants give a jungle air to the desks. The workforce is youthful and intense, with books and papers stacked to leave a channel of sight clear to their screens. The place has the feel of a university campus — some cutting edge institution, like I would imagine an austerity version of MIT to be. In terms of a gathering of different nationalities, this could be the United Nations. In terms of gender balance, women rule.

In November 2016, a delegation of eight eminent Chinese judges came on an inspection visit. Since their teens, these judges had been working up to 18 hours day, six days week, in what they see as being the service of the people. They found their way under the railway, squeezed through the metal security gate, and edged past the run of railway arches. Just inside the front door was the seminar room, where the judges sat in a circle with ClientEarth's senior team and discussed environmental decision making. Wei Wenchao, Deputy Chief Judge of the Environment and Resources Tribunal of the Supreme People's Court, listened carefully and spoke softly. 'It is very important that we visited your office,' he told James as the meeting ended, 'to see how much you have given up in order to serve the needs of the people.'

ClientEarth's tenancy of The Hothouse started with one glassed end of one upstairs floor. It has now spread along the whole floor, up to its galleries, and across the ground floor, and to meet with Alice I walk up stairs to a second floor conference room. I had passed that room just the week before. Its door was open because it was full, and I heard laughter spilling out. I looked in to see women from South Korea, the former Environment Minister and her entourage, excited to have dropped by for a morning seminar from James.

I set my recorder on the table and Alice breezes in. Sure, it's fine to record, she says. The recorder is less of a violation that it once might have been. Her first surprise on taking on the new role was having to deal with the media. 'Normally, as lawyers, you get to sit in your office and keep everything confidential, particularly litigation,' she says. 'We're taught not to talk about our cases and our theories.'

This is a sea change, or perhaps some tidal drift, since I started work on the book some years before. UK lawyers were initially very hesitant about claiming any sort of public profile. 'The challenge is to persuade legal minds that they are able to get change not only through behind the scenes advocacy with policy makers,' Tim Reid, ClientEarth's Director of Communications, told me, 'but also pile pressure on the policy makers by getting the public on your side to get what you are looking for.'

Other environmental groups have membership bases that are dependent on a high public profile. 'The environmental lobby group is hugely competitive,' Tim told me. 'The more we break through on air quality and pollution, the more our competitors, who are also our coalition partners, try to steal our thunder. Our challenge is that by using the law, we set ourselves apart from those organisations.'

To educate the public about using the law is one of the purposes listed in the charter of ClientEarth. Foundations tend to limit their funding to three or four year periods. The Esmée Fairbairn Foundation has stayed a core funder of ClientEarth from the start. Interestingly, the law group's ability and readiness to communicate is central to that partnership. 'What has impressed us most is not primarily ClientEarth's capacity to hold to account those who might jeopardise the natural world unlawfully,' said Caroline Mason, the Foundation's Chief Executive, 'though this has led to many significant victories. It is also in its willingness to inform, educate, and persuade those with the influence and the means to do the right thing and, by doing so, bring about the sustained changes in culture and attitudes that must happen in the face of the multiple threats to the planet.'

Alice reflects on this. 'It was obvious that ClientEarth was very much focused on the law,' is how she sees it. 'The mechanisms for change would be designed with the law first and foremost. If you want to make impactful change, it's got to be with the law. You also have to be able to communicate that in a way that will be taken seriously, rather than having a campaign hat on. Litigation trains you in how to construct an argument in the most persuasive way possible, pulling together the right evidence, and so it's about being persuasive.'

Alice was back home in London when news filtered through of the Paris agreement. All of the attending parties, 196 countries, had signed, with a limit of 2°C warming but a target of 1.5°C. 'I went to bed still refusing to believe it, woke up Sunday and it finally sunk in. I was speechless, exuberant all day, walking around in a daze. Then I got an email from Howard Covington.'

Howard Covington was a trustee of ClientEarth, a highly successful investment banker who had turned his focus to alerting the investment community to the economic risks of climate change. Paris had produced an amazing outcome, he told Alice. How do we push forward from this point?

Alice sat down with the challenge, and came up with her response: change the duties of company directors. A trio of factors guide directors in how to manage their business: the company's constitution, the Companies Act, and their expert knowledge. Alice took Shell's constitution down from her shelf. She drafted a new clause, and found where to insert it.

In the months before Paris, 2,043 corporate bodies made commitments to climate goals, and 1,200 signed the Paris Pledge for Action. Companies boosted their profile with green assurances, but how could they be held to account? Alice adopted a maxim for herself: 'governance not greenwash'. States have laws, but each company also has a law by which it is bound: its written constitution. The Paris Agreement's Article 2(1)(a) gives them the ideal model paragraph to insert. Alice wrote it out:

> The Company will undertake its business in a manner consistent with the objectives of the Paris Agreement (in so far as they relate to the Company's business), in particular, the objective to hold the increase in the global average temperature to well below 2C above pre-industrial levels and to pursue efforts to limit the temperature increase to 1.5C above pre-industrial levels.[7]

Leading barristers in the field agreed with the idea. Academics agreed. A group of CEOs was enthusiastic about it. A radical idea caught some immediate momentum. Plans rolled out to have some leading global companies in the field adopt the clause. Then shareholders might muster around and make it universal, to establish a mark by which consumers could trust a company's green credentials.

'It's amazing here to actually get paid to think ten hours a day just about solutions to climate change,' is Alice's take on 'the public interest law group difference'. 'I feel like the luckiest person in the world. It's absolute freedom, and to be surrounded by brilliant people who have the same interests. The constitution idea was born of four work streams: shareholder resolutions, fiduciary duties of investors, corporate reporting, and company directors' duties, and all the work that my team and I have done on them. Why would you think of something like that unless you had the opportunity of looking at the four constituent parts of it? You're in this melting pot of ideas and solutions.

'I've spent years of my life being really depressed about climate change, and pessimistic and very sad,' Alice concludes. 'Since starting here, I'm optimistic. If you're paid to look, there are solutions.'

And she hurries downstairs to look for some more.

Protestors held up banners outside Central Hall in Westminster. 'Shell NO!' they stated. 'Climate Justice Now!' A ship was in harbour in faraway Portland, Oregon, waiting for storms to abate so it could sail equipment out to the Chukchi Sea. Shell had a licence to drill the Arctic for oil, which was possible until ice sealed off the drill site for the winter. I snapped a photo of the laughing protestors and walked past them.

Polite guards at the door checked my documents and let me through. Shell had held its AGM in the Hague the day before. Now the company had come to London to speak to its shareholders. I am one such. Both Shell and BP send me occasional cheques for the likes of 37p, dividends from my ownership of a single share in each of their companies.

I had flexed my muscle by proxy at the recent Shell and BP AGMs. I wasn't alone. CCLA, a British investment firm, led a band of 55 institutional investors, including Jesuits in Britain and the Joseph Rowntree Foundation, in the filing of shareholder resolutions. Share Action, a charity focused on 'an investment system that is a force for good',[8] gave support, while ClientEarth's lawyers advised the group on how to navigate the strict requirements for resolutions set out in the Companies Act.

The resolution required the companies to submit reports on how they were adjusting their business model to the effects of 2°C global warming. The reports should detail any bonuses associated with climate harming activities, and show attempts to reduce emissions and invest in renewable energy. Investors could then decide whether to continue their investments.[9] The resolution passed with 98 per cent of the vote.

Shell's Chair, Chad Holliday, began the London meeting by saying how they were 'proud as a board to support the resolution'. The meeting rolled out in a well mannered way from there: shareholders lined up behind a pair of microphones in the aisles, while the executive team sat behind a table on the stage. Holliday took questions, alongside the Chief Financial Officer Simon Henry, and Shell's CEO, the Dutchman Ben van Beurden.

A public shareholder meeting is stage managed to show a board on its best behaviour, and anxieties had been allayed at the recent AGM. Still, I expected the occasion to be something of a heist, with activist investors ramming their issue home. Debate was in fact impassioned, but quite decorous. Shell pays high dividends. Several shareholders spoke as the inheritors of family funds invested generations ago. Shell had many live issues, including an ongoing corporate takeover of BG Group, so it was singular how the vast majority of questions concerned the environment. Activist shareholding clearly can raise ecological issues up the corporate agenda.

Ben van Beurden was quite fired up himself. Shell was 'one of the first companies that acknowledged climate change was a clear and present risk', he declared. No longer just an oil business, it had ramped up to be 'oil and gas'. 'Project Quest' would see a new plant sequester one million tonnes a year of CO_2 from the tar sands of Alberta, Canada. As a chemical engineer, he found the next step for carbon capture and storage to be particularly bracing — not just to store the stuff underground, but to find a viable economic use for it.

This was May 2015. The Paris talks were months ahead, but could be no more than a beginning. 'Paris is an important milestone,' he averred, 'but it is just a milestone. Targets will be seen as not enough, and we will need to do better. It's good to have targets, but it is more important to have policies that will lead to those targets.'

Chief among those targets was carbon pricing. Carbon capture and storage could go nowhere without a fair price on carbon. Currently, Shell was factoring $40 a tonne into the economics of future projects, and personally he felt that should be $60–80. 'Putting a real price on carbon is absolutely key,' he said, for it shifts power away from coal.

I headed for the aisle and waited in line to pose my question. Under Alan Rusbridger's final months as editor, *The Guardian* had been running a 'Keep It In The Ground' campaign, urging disinvestment from the fossil fuel industry. Separately, Howard Covington was using a series of papers and talks to raise the alarm about oil reserves as 'stranded assets': oil and coal would be seen as such bad actors in the climate change scenario that they could never be used, which posed huge risks to an investment portfolio. I asked, how were these campaigns going down at Shell HQ?

Not well, clearly. Van Beurden leaned forward so as to stare through the lights and straight at me. He spoke sharply, he admitted, because it meant a lot to him. A campaign such as 'Keep It In The Ground' ignores reality, he explained. It's a 'seductive argument' but 'a red herring'. It creates the illusion among a large part of the public that this approach offers a solution. Those who peddle such ideas are doing society a massive disservice. He doesn't have any grandchildren as yet, but even so it is doing a disservice to them too.

'I am prepared to lead a debate on this,' he concluded, 'because it is more dangerous to have a debate going on in a negative sense.'

Voila, I conjured exactly such a debate out of my pocket!

Well, not quite. I waited almost a year. The boards of Anglo American, Rio Tinto, and Glencore followed Shell and BP's example in the meantime, and supported their own 'strategic resilience' resolutions.[10] I printed out my notes, and took them to Howard Covington's home in London's South Kensington.

THE JUDGEMENT OF PARIS

As founder shareholder, director, and then CEO of New Star Asset
Management, Howard Covington ran 400 employees who managed £20
billion of assets. Cleverer still, he sold the company in 2009. He has the
intellectual clout to have become a fellow of the Institute of Physics, and
in 2015 the first Director of the UK's National Research Institute for
Data Science, the Alan Turing Institute.

In venues from the *Financial Times*, *Nature*, and the chinadialogue
website, Howard had been advocating the notion of fossil fuel reserves as
stranded assets. I watched him give a presentation at the London School
of Economics, with coal the villain of many a graph and bar chart. But
oil was in second place.

I challenged Howard. His whole stranded assets argument was
delusional, I suggested, reading from my Ben van Beurden notes. He
was doing society a massive disservice.

He twinged a smile. 'Mr van Beurden is talking his own book,' and
so of course the man was doing his job. BP's energy outlook of February
2016 shows demand for fossil fuels increasing around 1 per cent a year.
At about page 90, however, I would find an analysis of what happens if
fossil fuel demand peaks in the 2020s. Electric cars may take 20–30 per
cent of new vehicle sales by 2030. Since 50 per cent of oil demand is for
transport, replacement of petrol vehicles by electric would by itself cause
oil demand to peak. We had just seen a slight decrease in demand for oil
trigger a dramatic drop in its price. Once the oil price peaks in the 2020s,
the oil business will change out of recognition. A permanently low oil
price may send the global economy into a period of disarray.

'That's the real nub of the stranded asset argument,' he concluded.

Which brought me to the nub of my own quest. Howard Covington
was a man who reaped vast success from acquiring, merging, holding,
and selling companies, and was on a new mission. He had scanned the
options, and elected to work with ClientEarth. I settled onto the sofa of
the upstairs living room, soothed by the clean geometries of paintings

from the St Ives School that hang on the walls. My recorder flashed its red light, and I started to explore the meeting of the financier and the lawyer.

What use was a public interest environmental law group? Why seek out ClientEarth?

Howard spotted two different questions, and started with the second. As a child, Howard wanted to be a mathematical scientist of some sort, and in his answer I spot the mind of a physicist at work.[11] 'I wanted to meet James because I thought the little things I do could be magnified by the much greater things he does, and those in turn could be magnified by the rulings of judges.' There you have it: a fascination with scale, and how movement at the tiniest level affects the whole. 'We could go from little causes to quite big effects.'

So what's his big effect to be? Changing the global economy in six decades. That takes strategy, which might come from some intersection between switched on minds in Science, Economics, Finance, and the Law. It also takes the ability to care. If you can't care about the environment — and 'My erstwhile colleagues in the City [the financial district of London] almost without exception couldn't care less about climate change' — then care about your money.

It's hard to get these investors to think in 40 year horizons. Three hours is the norm, or maybe the end of the day, or the three months to the end of the quarter. Pension fund trustees have a 50 year horizon, but seldom think about the long term effects of climate change either. Nor do their advisers bring it up as an issue, because they don't want to frighten away their clients. Advisers and trustees form an alliance of the timid. Make it a regulation, trustees told Howard in their meetings, so that everyone is compelled to take climate change risk into account, and we would have no trouble in complying.

So that's what Howard aims to do: bring in the regulations. His City colleagues need the same regulatory spur. 'They are hard to persuade to change their mindset in any way other than through a change in regulation or legal judgement. James and his colleagues' view of the law is very much the view that investment bankers take. Within the parameters determined by the law, how can one structure things and go about developing a process that will lead to the answer you want? If ClientEarth can change the laws and regulations, then they can change the behaviour.'

Which returned us to the original question: what use is a public interest environmental law group?

Howard had his answer ready. 'It's very hard nowadays for individuals to stand up for themselves, because the cost of bringing a legal action is enormous. It's only a public interest law firm that can really stand up to governments or large corporations. We will just see more of what ClientEarth does in the future, not less.'

Had he considered working with any other of the green groups? I asked. He stared at me long enough to make me feel stupid. What other green groups focused on the law?

The big campaigning groups saw law and lawyers as being the servants of their campaigns, I countered.

'Ha!' he said, for it made sense. This notion of a public interest law group came from America. Howard liked Americans: they generally believe they can change things even when there are major obstacles. 'In the UK, investment banks lead transactions, and the lawyers support them; in the US, the lawyers lead the transactions. The idea of lawyers in the lead is a transatlantic idea. It's very usual for the lawyers in the US to be at the head of the table.

'All environmentalists are fighting against the odds. Trying to change how $70 trillion of economic product is made each year is a very big deal. You have to be very committed and optimistic to think you can make a difference. What I know is that law is a very powerful way of changing this. Not only is optimism a characteristic of our cousins across the water, but also law is a characteristic. An optimistic application of law is a US trait.'

Our Saturday flight to New York never took off. An all time record 775 millimetres of snow fell on John F. Kennedy Airport instead. The state had a lockdown on all transport as the winter blizzard of January 2016 stormed along the Atlantic coast.

Even so, January 2016 was globally the warmest in 137 years of record keeping, 1.04°C above the 20th century average.[12] The effects were even more extreme at high latitudes, with areas of the Arctic Ocean 6°C above average, which prompted in turn a new low for Arctic sea ice. 'We've got this huge El Niño out there, we have the warm blob in the northeast Pacific, the cool blob in the Atlantic, and this ridiculously warm Arctic,' reported Jennifer Francis, a climate researcher at Rutgers University. 'I think this winter is going to get studied like crazy, for quite a while. It's a very interesting time.'[13]

We land two days later. Highways away from JFK airport are stuffed with traffic. The pavements of Manhattan are clear. To leave them and cross the roads, people have kicked and trodden narrow passages through the ploughed snow piles. Trucks soon come to scoop these piles away. The storm killed six in New York, but otherwise the city is dusting itself down and marching on.

I slip fancy shoes into my backpack and put on boots to walk through Central Park. Thick snow stretches for untrammelled acres here. I search for a snowman but find none. It seems New York snow is to be dealt with and not befriended.

In the lobby of a fancy apartment block beside the Park, I swap my boots for shoes. James has brought a small team from ClientEarth; Frances Beinecke, former President of NRDC and now a ClientEarth trustee, has flown in from giving a speech in San Francisco; and the evening's host, Liesel Pritzker Simmons, has come down from Boston to her New York home.

This is the launch event for ClientEarth's New York fundraising office. The strong tradition of public interest law in the US has settled the concept in the nation's psyche. Liesel is excited by ClientEarth's expansion into China and Africa. 'The need is dire and the market's big,' she states. 'It could really do with a lot of players that are working in complementary fashion. I'm hoping that an American audience is going to get excited about this global approach.'

Waiters have canapés and glasses set on trays, and guests are due. Liesel still has to change. I want half an hour for an interview. Remarkably, she tucks her feet beneath her on a sofa in her library and settles in.

I'm a little nervous. Interviews always have an edge. I've flown thousands of kilometres for this one, but that's not the distance that matters. It's the metre across the coffee table. Both Liesel and I have to step out of our lives for the interview to work.

Why the nerves? Liesel found world fame before hitting her teens, playing the lead role in *The Little Princess* and Harrison Ford's daughter in *Air Force One*. I know world famous actors and can cope with that. Liesel is also fabulously wealthy. When your family owns the Hyatt Hotel chain, that's what happens to you. *Forbes* estimates her fortune at $600 million. That's top of the wealth league, but I have known some other fabulously rich people too.

So why those nerves?

Because she's smart. Liesel stepped back from being a celebrity because life has to mean more than that. She chooses to use her fortune as an agent of change. High powered advisers surround her. 'Hire people smarter than you,' is her motto. 'Listen to advice from people who you respect, and know what you don't know.' These advisers guard her well.

James first met her at a conference of some 300 social impact investors in Lugano, Switzerland. There was a computer dating service where you could log on and ask for a 15 minute meeting with any conference participant. Most everyone wanted to meet with Liesel and her husband, Ian Simmons. James asked and heard nothing back.

The conference organisers had asked for a presentation from James, which he duly delivered. He pointed out how law had brought about major victories in the battle to save the environment, and then switched to the value of his particular delivery mechanism. His public interest environmental law group cost only 10 per cent of what an international law firm would charge for the same work.

Liesel couldn't come to this talk herself, so she sent her accountant. He reported back that James's work was the most high leverage investment on offer at the conference. Duly vetted, James got his meeting.

Liesel doesn't suffer fools, gladly or not.

And here I am. On the matter of the law meeting with investment to drive an immediate change to the planetary economic structure, I'm out of my league. Yet for me, it's the big shadow left over this book.

The first big shadow was China. What did it matter how far the rest of the world corrected its behaviour if China's coal fired power stations belched ever more carbon? The question is still alive, but I trust China's ambitions for an ecological civilisation. I see how lawyers are being encouraged to install and enforce the appropriate regulatory structures.

The second big shadow was Africa. The continent is aching to develop. It is desperate for constant electricity supplies, which will ramp up carbon emissions. Its forests are liable to be chopped and burned. Funding to ensure sustainable development is liable to be swallowed up by a corrupt leadership structure. Corruption remains a Serengeti sized cloud. Yet I had seen how civil society in Africa is empowered by learning how to use environmental law as a tool. Law levels the playing field. I watched citizens discover their rights and apply them. Ecological crises can give rise to individual despair: 'What can I do? I can cycle and recycle while species die and seas rise.' In Africa, I saw how law was an antidote to despair.

And so I've come to New York to face down the third shadow: that gloomy disempowering feeling that individuals are powerless in the face of corporate structures vested in short term profits from unsustainable practices. As temperatures rise, the wealth gap increases, the top 1 per cent over the rest. Here I am in the Manhattan home of one of the 1 per cent, with more of them due for a fundraising dinner any minute, on an island anchored by Wall Street.

You magnify something so as to see it more clearly. Liesel's wealth is an insane magnification of most people's. How does she align it with environmental law to achieve an effect? What can the rest of us learn from that, so we spring from hopeless bystanders to ecological champions?

Those are big and unreasonable questions. I have half an hour to get the answers I need. Hence I'm nervous.

It turns out Liesel is nervous too. Interviews are intrusive, and she seldom grants them. The more I introduce myself, the more I make her want to walk away.

Off we go.

Liesel met her husband, Ian, at a meeting of global philanthropists. Together they formed Blue Haven Initiative, a single family office dedicated to social impact investing. 'Our mission and strategy,' Liesel explains, and I appreciate the combination of both nouns, 'is to find companies that are market rate for profit but their core business provides some kind of solution that has measurable environmental and social benefit as well. There are lots of those companies out there. We invest in them directly through funds, and through public equity and public debt vehicles as well. That's what our core business is.'

The nature of the game is social impact investing. Blue Haven prefers the term 'values aligned investing' and is a leader in the field. To give some sense of their vision, one $50 million slice of their fund is directed towards early stage companies in sub-Saharan Africa. Wherever they invest, the companies must have a strong set of ideals in place. Blue Haven provides capital, and expects profit in return. In that way, they become a model for alternative for-profit investment.

And for that to work, values aligned investing cannot be seen as a high risk strategy. On the contrary, it is climate change that needs to be seen as a risk.

In 2014, the billionaire investors Tom Steyer, Michael Bloomberg, and Henry Paulson issued a report, *Risky Business: The Economic Risks of Climate Change in the United States.* Liesel saw a movement begin to swing her way.

'People are starting to wake up to the fact that if you're an investor and you care about environmental risk, maybe you're just risk averse, you're not just a tree hugger,' Liesel says, and then dramatises the scene for me. 'To other investors I say, "Oh my God, you're not looking at environmental risks? Wow. OK. I guess you guys do the bare minimum of work. That would never pass muster in our shop." When you frame it that way, you get a lot further in the discussion than if you say "Save the Whales". We're very risk averse. We're very conservative. It's not a false argument; it's a true one. I am starting to see that shift happen. Every year, we have the hottest year on record. Every year, we start seeing commodity prices swinging wildly. It introduces a lot of volatility. As an investor, I don't like that.'

'Investor' is the self-definition. I wonder about philanthropy. Business is about getting a return on investment, and yet Liesel has donated to ClientEarth. Find an organisation to trust and then give to core funding: this is her preferred philanthropic mode. Agree on budgets, set milestones, stay in touch. She is supporting this fundraising office in New York. There's no obvious bottom line in that. What's the logic?

'If I ever step out of line as an investor I want to get smacked,' Liesel says, to my surprise. 'I want to get put back in line. To me, investing in ClientEarth is not just a do-good thing for the planet, I want to help fund those institutions that create transparency, that hold me as an investor and any of the things I invest in accountable to rules and regulations. It's exactly the role of philanthropic capital to help make systems better for holding people accountable.

'Traditionally, my only point of leverage as an investor, particularly with a company that I am a minority shareholder of, is to do proxy voting and shareholder advocacy. This is another hammer. I get excited about choosing different angles to get at the same goal, which is "How can we make the planet better, greener, but not necessarily stifle business in the process?" I don't think it's one or the other; I think it's both. But you've got to have good rules, and those rules have to be enforced. The work that ClientEarth is doing is fundamental in creating structure to what we do already on the investor side.'

My nerves have eased. Liesel too is relaxed. Her crossed legs are not just a casual posture. She vanished to India as a young adult and spent time teaching yoga and breathing techniques to drug addicts. That yogic poise is on display as we talk.

Funding a law group to ensure good governance so your sustainable investments bear fruit makes sense. I like the logic that entwines giving and investing, philanthropy and business. It's a new and useful insight for me. Yet I want more.

Snow has fallen. Liesel could be swanning around in a ski resort somewhere. She could be blinking out the window at a landscape. She could salve any conscience by giving to regular good causes. Instead, she flew through awful weather to give a day to launching the US arm of a charity. She's happy to talk to me. She's as cheerful as Spring. I want a dash of that bright sense of purpose. That's what I want.

So, what makes an impact investor and philanthropist buzz? I ask. What's the secret?

'I'm one of the luckiest people I know in terms of winning the birth lottery,' Liesel begins. 'I've access to lots of resources and access to a lot of choice as a result of that. The way that I was raised and what I have learned from both my mother and father is, that's not a given. And so how can I use the position I have been given by birth to help other people?'

That's the question posed to Liesel as a child. How do you use the position you have been given to help others? It's a question you can apply at any station in life. What would it mean for me? Kindly, she gives me an answer.

'There are a lot of different ways to do that, and a lot of ways people have chosen to. Some people make their career about direct service, and that's noble and brave and wonderful. Given my access to resources, I thought that the highest leverage thing I could do was to pay attention to those resources and what they were doing in the world. And hopefully inspire other people to pay attention to those things as well. Sometimes that takes the form of philanthropy, sometimes that takes the form of impact investing, and sometimes that means getting involved with organisations I really care about. For me, it always comes back to the veil of ignorance. What is, is not necessarily fair, so don't get too comfortable about it. You can't change the world on your own, but you have to move with people who are trying to.'

She laughs. I laugh along. I have what I need. The voices are loud in the corridor. Preparations for the party have gathered full steam. James is set to tell a room full of strangers why his work needs their help. They all want to change the world, to help it mend. I pick up my glass and go to move with them.

Conclusion

James Thornton

Law suffuses our lives from the public scale to the intimate. Since tribal times, whenever two or more gather, we make rules.

Birds have songs and we have laws. They define us.

Our laws often show the best side of us. In the West, we have seen a growing social enlightenment. Slavery was banned and women given equal rights. Gays are in the process of achieving equal rights. Laws protecting nature, the environment, and human health have been passed.

This new corpus of law, starting in earnest in the 1970s and continuing today around the world, has a special relationship to our better side. Clean air and water, forests, rivers, the ocean, and all wild animals are often referred to as a commons. In 1968, ecologist Garrett Hardin wrote a seminal article called 'The Tragedy of the Commons'.[1] He made clear that without protection, the commons will be used by the selfish. Everyone will have to compete to get a share, or do without, leading to destruction of the common resource.

Hardin's solution is 'mutual coercion mutually agreed upon'. This is a fair definition of how we make law in the modern world.

The Code of Hammurabi is from first dynasty Babylon. In Hammurabi's day, the king made the law. Today, we agree on the provisions of law in legislatures. In the case of legislation that protects nature, the environment, and human health, law protects the commons. This could mean restrictions on dirty air, of the kind we are enforcing against the UK and other governments. It could mean a regime to decide how to leave enough fish in the sea, as in the amendments we worked on for the EU's Common Fisheries Policy. It could mean that the people who depend on forests share in the benefits when trees are harvested, as we worked on in Gabon.

Turn now to the Paris Agreement, the greatest of projects to protect the commons. Here, the commons is the global atmosphere. Keeping it within safe limits is a precondition of a good continued life. To reach this agreement was one of the most difficult milestones to achieve in history.

Now the question is: what does the Paris Agreement really mean?

First, the Paris Agreement has enormous psychological value. It tells the story that humanity can come together to deal with our number one global problem. Before the Agreement, this was in doubt. Now there is a way forward.

Second, the Paris Agreement works as a global framework. It requires countries to go home and figure out how and by what amounts they will reduce emissions. It has legally binding provisions on monitoring and reporting emissions. With luck, this will lead to a virtuous competition in reductions.

Third, each country has to come up with its Nationally Determined Contribution, the amount of greenhouse gases it is going to eliminate. Here is where the Paris Agreement will succeed or fail. It is all about implementation and enforcement. Will countries go home and put in place real reductions, then make them legally binding? By legally binding, I mean something specific, which you will be familiar with from this book. Countries cannot sue each other under the Agreement, nor is there any global enforcement body. The enforcement question is simple: will countries allow citizens to haul them before their courts if they fail to make the promised reductions?

If countries do allow citizens to be involved in this way, it means the reductions are likely to be real. If they are only promised in policy, they amount to so much hand waving. Unless policy becomes law which in turn becomes enforceable by citizens, it is meaningless. It can be changed or ignored any time, and often is, after the next election.

The Paris Agreement will work if the reductions country by country are significant and enforceable. Trillions of dollars in clean infrastructure needs to be built in the coming decades. If countries whimsically change policies, investors will not gain the confidence they need to fast forward global society into a carbon neutral future.

Legally binding reductions are needed to release a torrent of investment, so this third level is crucial.

Fourth, moving to the company level, something else is needed too. Companies and investors need to understand they have a duty to move away from bad carbon investments towards a carbon neutral future. The investor and company duties derive from understanding the systemic risk to value that climate change poses, as pointed out in the previous chapter. Actions in the right direction of travel are supported by Paris, by the national reductions, and by the overall story that it is time to recognise climate risk and deal with it.

Fifth and finally, at the individual level, behaviour change is needed. Here again, story is important. What does my life mean in the context of climate change? If action is happening at every level, global, national, and corporate, then individual action no longer feels lonely. We will no longer feel isolated and powerless. Instead, we become part of a community of activity all focused on the common good.

This connection of our individual action with the common good allows us to rise above selfish needs into a feeling of shared responsibility, which generates meaning and purpose. That we are all in this together is a feeling that comes in wartime, but it also arises from working together on a common problem.

The promise of Paris is that it can mobilise action on the global, national, corporate, and individual scale. These could all cohere into a shared experience of caring for the world and saving the future.

Saving the future by acting in the present is what we need to do. Is this really possible?

Let us go back to the heart of where law comes from. Law emerges from the story a people has of itself. The Code of Hammurabi does not speak of the rights of women and minorities. The codes of many African countries still enable the extractive economies imposed by colonial powers, with African leaders now sitting in the colonial seat of power. As understanding deepens, the laws change to reflect the shift.

To see climate threat as shared is a fundamental shift. The law now needs to shape a set of directions for how to act. For the first time, the story and the directions need to be effective at every level of scale. Specificity is important. Vague generalities cannot be implemented and cannot be enforced.

It is worth repeating: if you pass a law and do not enforce it, you in effect authorise the behaviour you sought to prohibit. Obligations that are detailed enough to be enforceable must reside in the law. Then they must be enforced.

Every time our story evolves and the law captures it, requiring new behaviour, enforcement becomes an issue. In the civil rights era in America, it took years of enforcement before the change was real. The enforcement of the civil rights laws was done both by government and citizens. Importantly, those changes have not been fully realised even yet. Black people are still fighting for full equality and still need to. The same is true for women and for the LGBT+ community. When the cultural story changes, and the law requires new ways of behaving, it takes decades for all the blockages embedded in the system to be removed.

This lag in behaviour change is evident in the environmental arena. Modern environmental laws, as is described early in this book, got their start in America in the 1970s. Nixon signed them and was happy to do so. Then there was a backlash. By the time Reagan came to power, his government was determined to subvert the environmental law.

In Europe, the environmental laws followed about a decade after those in America, and there was also enthusiasm. By now, there is backlash of entrenched interests who do not want the fundamental changes that taking care of the environment and people's health requires. The UK government is doing what it can to undermine the rule of law for the environment, as the Reagan administration did. Each European country has its own problems.

China is where I find much encouragement these days. While their problems are severe, the intention to address them is real. From the grassroots to the top, they are saying the same thing: we need to end pollution. We need to rebalance growth with protection of ecosystems. The Chinese are pragmatic and systematic. They are studying the best global examples by sending their experts out and by inviting experts in. It is a sign of genuine openness that high Chinese officials are willing to take advice from the CEO of an environmental organisation like me.

The Chinese are also consciously working on a new cultural story. By making the concept of Ecological Civilisation the organising principle of their actions, they are trying out a story that may carry deep change forward.

What I mean here by a story is the foundational myth of the culture, the ground norms that determine what people take as valuable. In medieval Europe, Christianity shaped what people spent their time and thought on. In contemporary Europe, the myth, shared with America, is neoliberal capitalism. It elevates private gain as the focus of human life.

Our current myth led us to globalisation. While there are benefits from globalisation, particularly for the poorest, there are growing concerns that globalisation is eating away at the civil virtues that create that space of trust in which we come together to work and live and love. Globalisation has greatly concentrated wealth at the top, and left the working class in the industrialised world feeling left behind. Their real income has stagnated, their jobs have become less secure. They sense the West drifting towards plutocracy, where Western democracies favour the super wealthy while pretending to listen to the people. A responsive anger has arisen, leading to populist candidates on both left and right having success that was unimaginable a few years ago.

The direction of political travel in Europe and the USA is not reassuring. On the one hand, we see increasingly entrenched corporate interests. On the other hand, we observe the rise of populist demagogues who would happily show a fascist face once in power. Neither option is acceptable. Avoiding them needs a renewal of democracy.

Such a renewal happened after World War II and could happen again. Fighting for the right of citizens to a healthy environment is a key part of this renewal of democracy. We are constantly arguing for the rights of citizens to be protected from unchecked corporate behaviour.

Making corporations into responsible civil actors is the basis of laws that protect environment and human health. Ensuring that citizens have access to information, that they have the right to participate in decision making in matters that affect them, and that they have the right to go to court to defend their rights — all these are central to renewing democracy in our time. To renew democracy, empower citizens, make sure that economic policies favour them, and ensure they have a healthy world for the long term. Protecting health and the environment, and empowering citizens, can be the fulcrum around which the West renews its democratic values.

In China, I was struck by hearing from senior Chinese officials that they were proud of raising more people from poverty than anyone else in history, but that they got the balance wrong. They drew down the environment too much and now would do whatever is necessary to restore the balance.

This is a wonderfully clear analysis. And the story of Ecological Civilisation is meant to sustain the effort to shift the balance.

How do we make this vision real?

Here we run into our own biology. Our brains are not hardwired to see long invisible problems like climate change. They are well designed to see the immediate threat. So terrorism takes centre stage, or the rise of demagogues. Refugees capture the attention. Thousands, then millions, will be climate refugees. This will become a preoccupation, and will give rise to reactionaries, with the temptation to build walls and close borders. There will be regional conflicts, caused or worsened by climate change.

Unless we address the long game well, nothing else we do will matter. Civilisation itself is at issue. I was pleased recently to meet with Jeremy Grantham, a leading investor who focuses on climate change. As he puts it, what we can anticipate in a few generations is rolling collapse.[2] The tropics will suffer first. Centres of culture will survive for some time in the cooler regions. It is a disturbing picture when you study it.

Martin Rees, Britain's Astronomer Royal, studies existential threats to civilisation. In a book called *Our Final Century*,[3] he looks at the problems facing us, including climate change. Rees believes we are in a bottleneck of dangers this century. He concludes that we have about a 50 per cent chance of survival, and need long term thinking to do so.

The great Harvard biologist E.O. Wilson recently published *Half-Earth*.[4]

He too sees us as being in a bottleneck, in his case pointing to our loss of biodiversity. Our economic activity has ignited the sixth great extinction, in which we now live. Wilson also finds a 50 per cent solution, in this case protecting half the planet for the natural world, so that it can continue to nurture humanity.

We could solve the climate crisis as a matter of physics and chemistry yet stay on our killing spree towards nature. This would leave us with civilisation but on an impoverished planet. No one in their right mind would think that wiping out half of all species by the end of the century would put us on the planet we want to live on, but that will come from business as usual.

This dual focus on nature and climate is why I set up ClientEarth the way I did. The environmental groups founded in the 1970s work on a broad range of issues. Not so those founded more recently. Younger groups, even large ones, specialise. They work on energy or fish, toxics or birds. I had a different vision.

Because the problems are interconnected, I set up the first environmental organisation in recent decades that focuses on all the important issues. We have a broad focus because only a holistic understanding, which arises from working on all the key problems in their interconnection, can deliver the solutions we need.

Nature, as mentioned earlier, speaks to us in the grammar of science. So write laws grounded in the science. Make them reasonable, so they can be accepted by the regulated industries, and understood by the regulators. Implementation is not possible if the regulators and industry cannot understand or accept the regime a law sets up. When they do, respect for the law emerges.

Enforcing the laws is also required for their respect. We will therefore need to keep enforcing the law, wherever we work, and help other citizens do so.

If we did succeed in enforcing all laws protecting the environment and human health, however, it would still not be enough. Environmental law is young. Our understanding of the science is rapidly growing.

Environmental law needs to evolve as our understanding evolves.

If we are to solve the existential threats to biodiversity and civilisation, we need to build a newly comprehensive environmental law. Call it Environmental Law 2.0, and describe it this way: the set of laws that, if implemented and enforced, would keep climate within safe limits, protect biodiversity in a rich state, and protect human health.

It will not be easy. We need to refashion our agriculture, industries, and transport so they are carbon neutral and support the web of life. To do this, the money will have to follow the new story we learn to tell. But the money always does follow the story. That is what money does. The story of a culture shapes the concerns and creates opportunity. It determines where investment flows. If we need $90 trillion investment over the next 15 years, then we need to create the certainty that opens the opportunities investors need.

To make all this happen, we need the motivating story, the myth that will focus our attention. Only such a story will keep us focused on the long term as distractions grow. I hope that our sisters and brothers in China have found in Ecological Civilisation a story that will touch our deep nature to motivate our actions.

We will need detailed rulebooks of how to act appropriately. Understanding how climate change risk comes into the duty of every actor in the investment chain will help. When climate change risk, left unmanaged, is a personal risk to professional investors, investment will move the right way.

There are two bottlenecks we have to get through: climate change and the extinction crisis. The way forward is a twin path. First is learning to tell a new story that focuses us. Second is embodying this story in rules that give us clarity on how to act when we invest, when we legislate, when we enforce, and when we as actors take all the other decisions that shape our cultures.

This new story and such a tapestry of guidance will let us each contribute to the growth of culture beyond the bottlenecks, so that people and the rest of life flourish into the far future.

I have no doubt that we can save the future by present action. We are capable of acting wisely. Wisdom and altruism are as much a part of our genetic inheritance as greed and aggression. The lesson of the Paris Agreement is that all nations share the beginning of the new story we need. Let us work together and realise the dream.

Postscript:
Hope in the Time of Trump

James Thornton

Since this book was written, Donald Trump has assumed power. If you are concerned about the environment, nuclear peace, healthcare, corruption, and much else, Trump's malign actions can be a source of despair. By announcing that he will take the US out of the Paris Agreement and by playing at nuclear brinkmanship, he has declared war against the future.

He has put people in charge of US agencies who seek to undermine their missions. Scott Pruitt, his head of the Environmental Protection Agency (EPA), is working systematically to dismantle any American response to climate change, while promoting coal. He has requested deep cuts to the EPA's budget. He intends to fire a lot of its staff. He bans agency scientists from speaking about climate change, as if removing climate change from the dictionary will make it go away.

When I was a young lawyer, Ronald Reagan came into office. Reagan too had a radical environmental agenda. He appointed Anne Gorsuch to run the EPA with the mission of bringing it to its knees.

While Reagan managed to do damage, it was temporary damage. As narrated in the opening of this book, I worked hard against Reagan. Many others did too. In the end, he failed to do the damage to the environment that he intended.

I believe much the same will happen with Trump. He will fail to damage the planet as much as he hopes.

When it comes to climate change, he may take America out of the Paris Agreement. The way the agreement works though, leaving will not be effective till the day after the next presidential election. If America does leave, it will be the only country in the world not party to the Agreement. On the other hand, if he fails to win re-election, or has a moment of temporary sanity, America may stay in. An overwhelming majority of the American people, according to polls, want to stay in.

But it may not matter much whether America stays in. Fourteen states and counting, including New York and California, have created the US Climate Alliance, pledging to follow the Agreement. California, the 6th largest economy in the world, has already signed a bilateral deal with the Chinese to reduce emissions. These states represent much of the economy of America, so America may, in effect, stay in.

If America stays in by proxy because states reduce their emissions, there will be a pleasant twist on conservative doctrine. Trump and American conservatives have long said things should be left to the states, with the federal government dismantled. Now Trump is flexing federal muscle to pull America out of its global responsibility on climate change, and it is the states who are acting. An intellectually consistent conservative could only applaud the states's initiative and their assertion of the right to protect their citizens from climate change.

Reagan's war against the planet was one he lost. Now Trump has declared war against the planet. I believe he too will lose. There are four pillars to my belief.

First, environmental protection is much more deeply embedded in American life than it was under Reagan. When Reagan came to power, some of the great environmental laws that Nixon signed were not even a decade old. The agency set up to keep the air and water clean, and protect people from toxics, was also new. Now the laws and the agencies have been doing their work for almost 50 years. They are much more embedded in the landscape than they were then, with deeper roots, harder to pull up.

Second, citizens who oppose Trump's war against the future are stronger. When I worked at NRDC during the Reagan years, we were about 100 people. There are around 600 there now, and similar growth has happened at the other environmental groups. So there is an expert and disciplined army to oppose Trump's negligence and destruction. And it is working. Lawsuits are slowing down and stopping some of the worst excesses. Judges are holding the government to the law. This will keep happening. Sweeping illegal changes won't work.

Third, by failing to protect human health and the environment, Trump is thwarting the will of the people. Like all demagogues, he likes to talk about the will of the people, without caring to listen. What the people want is clean air and water, and a future for their children that is not rendered dystopian by climate change. This is as clear in the polls as the people's disapproval of Trump. Helpful too is that Trump's dream of bringing back coal is a fantasy. No sound market analyst thinks that coal can return. A consistent conservative would laud the market for moving towards clean energy.

The fourth pillar of my hope lies outside the United States. America elected a property developer and game show host with no relevant experience to the office of president. In China, a man like Xi Jinping comes to power only after decades of taking on ever more demanding roles. While Trump tweets, Xi has consolidated power. He has become the most powerful leader since Mao. His political philosophy is now enshrined in the Chinese Constitution.

That makes what he thinks of global importance. While Trump works to pull America behind its border, cut trade ties, reduce foreign aid, and abrogate treaties, China is doing the opposite. After the Second World War, America engaged in the Marshall Plan, pouring investment into Europe to rebuild it. The Plan was designed to build peace and trading partners.

China is now engaging in something similar. In developing countries around the world, China is building infrastructure. The plan is to increase the economic capacity of the partner countries, while building allies and markets for China. It is important to know that this project, the Belt and Road initiative, is on a much grander scale that the Marshall Plan. So while American pulls its head into its shell, China positions itself to become a global leader.

What is this global leader saying about the environment? Xi gave a speech a few weeks ago at the 19th Party Congress that consolidated his authority. One of the subjects he covered was the environment. Reading his remarks, they are a more comprehensive statement of dedication to address climate change and environmental problems than the leader of any other important country.

His opening was:

Man and nature form a community of life; we, as human beings, must respect nature, follow its ways, and protect it. Only by observing the laws of nature can mankind avoid costly blunders in its exploitation. Any harm we inflict on nature will eventually return to haunt us. This is a reality we have to face.

The modernisation that we pursue is one characterised by harmonious coexistence between man and nature. In addition to creating more material and cultural wealth to meet people's ever-increasing needs for a better life, we need also to provide more quality ecological goods to meet people's ever-growing demands for a beautiful environment. We should, acting on the principles of prioritising resource conservation and environmental protection and letting nature restore itself, develop spatial layouts, industrial structures, and ways of work and life that help conserve resources and protect the environment. With this, we can restore the serenity, harmony, and beauty of nature.

He went on to discuss green development, and how to address serious environmental problems, intensify the protection of ecosystems, and take leadership in building an ecological civilisation.

Here we have a comprehensive understanding of the need for working with nature. What will matter, of course, is how far his vision is translated into action. What gives me hope is the clear understanding the Chinese leaders have that it is in their own self-interest, and that of the country, for serious environmental action in concerted fashion.

Whatever you may think about China's form of government, or their record in human rights, I believe that when it comes to the environment, the political will to do the right thing is there, which is what makes it so exciting to work in China.

Trump has declared himself the enemy of the future. He wants to hide behind his borders and wreak havoc. That havoc will be contained by the law, the courts, and organised groups of citizens who know better.

The rest of the world, including an ascendant China, sees the need to work together for a common future. This gives me boundless hope.

Berlin, November 2017

Acknowledgements

The idea for this book came from Winsome McIntosh. Subsequent funding from the McIntosh Foundation provided two years of research time for Martin and travel costs administered through the University of Hull. The McIntosh Foundation pioneered funding of public interest environmental law groups in the United States. Winsome sensed that Europe had need of this book so as to learn how strategic use of law could help protect the environment. She asked for a book that would reach the mainstream, and is particularly keen that philanthropists come to see how use of the law can leverage their money, but granted us interviews and a free, open, and totally trusting hand.

Michael McIntosh supported the project as President of the McIntosh Foundation, and as the first interviewee. His vision set in place the story.

The law group ClientEarth features heavily in the story. Not everyone welcomes the intrusion of a biographer into their work and lives, and we have respected the wishes of those lawyers who preferred not to contribute or to have their story told. Big thanks to those staff members from ClientEarth who accepted and sometimes embraced the interview process: Clement Akapame, Dimitri de Boer, Alice Garton, Karla Hill, Maria Kleis-Walravens, Ludwig Krämer, Sandy Luk, Tim Reid, Małgorzata 'Gosia' Smolak, Marcin Stoczkiewicz, and Jozef Weyns.

Others who were generous with time and interviews that gave this book much of its substance were Laurens Ankersmit, Glen Asomaning,

Richard Benyon, Simon Birkett, Chris Butler-Stroud, Howard Covington, Cory Edelman, Connie Hedegaard, Harvey Jones, Aniela Kamińska, Maria Kenig-Witkowska, Elvis Kuudaar, Elena Visnar Malinovska, Caroline Mason, Samuel Mawutor, Kwame Mensah, Phil Michaels, John Mulligan, Samuel Naawerebagr, Nana Tawiah Okyir, Matt Phillips, Liesel Pritzker Simmons, Osofo Kwasi Dankana Quarm, Tim Robbins, Kristina Sabova, Danyal Sattar, Mustapha Seidu, Mariusz Sledź, Jean-Luc Solandt, Jan Šrytr, and Monica Stefańska. They shared stories and expertise from lifetimes of commitment in support of the natural world.

Małgorzata Smolak set up the trip to Północ, and sped Martin there and back by car. She was a cheering and resourceful guide, translator, and cultural interpreter throughout the whole adventure. We thank Aniela Kamińska for her poem, and Gosia Thornton for its translation. For the Ghana trip, Jozef Weyns worked hard in advance with interviews and materials so Martin stood some chance of understanding what civil society means in its African context. Josef arranged the whole trip, welcomed Martin into his workshops, and was a terrific companion throughout.

We are deeply grateful to Brian Eno for his foreword. Brian has been a ClientEarth trustee, friend, companion in thinking, and generous supporter right from the start, and, since he is so attuned to the Zeitgeist, his involvement gives us hope that what we are doing will actually connect with the culture.

Of course, we are liable for any errors or misconceptions, but are grateful for expert edits from Anaïs Berthier, Simon Birkett, Vito Buonsante, Dimitri de Boer, Alice Garton, Karla Hill, Sandy Luk, Tim Reid, Małgorzata Smolak, Marcin Stoczkiewicz, and Jozef Weyns.

Bill Swainson, then of Bloomsbury, helped reconfigure this book in its early stages. The project did very well to come into the hands of Patrick Walsh, an agent who cares acutely both for books and for the protection of the living planet. He helped steer the book into being and then into the hands of Philip Gwyn Jones, an editor and publisher who also values good books alongside their power to effect change in the world. He turned the project into this book and so gave it a readership. So we're immensely glad of Patrick and Philip, and of you readers. The world is heaped with books. Thanks for giving this one some of your time. If it spurs any awareness or action in your life, we're grateful for that too.

Notes

Introduction

1   To watch the speech: https://live.ft.com/Events/2016/FT-Innovative-Lawyers-Awards-Europe-2016?=&v=5164321407001 Accessed October 30, 2016

2   Sarah Murray, 'Law Firms Look Less Firm', *Innovative Lawyers 2016, Financial Times*, October 6, 2016, p. 17

3   The phrase 'Sue the Bastards' was wielded by Victor Yannacone, the lawyer at the heart of founding the Environmental Defense Fund (EDF). According to Robert Gottlieb, the phrase 'reflected the sense of implacable opposition associated with direct action forms of environmental protest while placing such protest in the context of litigation'. Later leaders of EDF would claim this 'worked well in the pre-NEPA, pre-environmental policy system days of passion and advocacy ... But by 1970 that approach was seen as increasingly problematic by key organization figures.' Robert Gottlieb, *Forcing the Spring: The Transformation of the American Environmental Movement*, Island Press, Washington, DC, 1993, p. 138

1. The Voice of Many Waters

1   Observations by George Percy, archived at http://www.americanjourneys.org/pdf/AJ-073.pdf

2   For a stirring tale of early Chesapeake Bay and those oyster wars, see Steve Nicholls, *Paradise Found*, University of Chicago Press, Chicago, 2009

3   H.L. Mencken, *Happy Days 1880–1892*, Johns Hopkins University Press, Baltimore, 1996, p. 55

4   Victoria Churchville, 'The Poisoning of Chesapeake Bay: Pollution Permit System Abused by Industry, Sewage Plants', *The Washington Post*, June 1, 1986, http://www.washingtonpost.com/wp-dyn/content/article/2008/12/02/AR2008120201990_pf.html Accessed November 29, 2016

5   Russell E. Train, 'The Environmental Record of the Nixon Administration', *Presidential Studies Quarterly*, vol. 26, no. 1 (Winter 1996), p. 185

6   ibid., p. 195

7   Gregory R. Signer, 'Do We Make a Difference?', *Natural Resources & Environment*, vol. 19, no. 1 (Summer 2004), p. 74

8   Matthew D. Zinn, 'Policing Environmental Regulatory Enforcement: Cooperation, Capture, and Citizen Suits (part 2)', *Stanford Environmental Law Journal*, vol. 21, no. 1 (January 2002), p. 81, http://web2.law.buffalo.edu/faculty/meidinger/561/materials/Zinn2.pdf Accessed March 26, 2015

9   Anne Burford, *Are You Tough Enough?*, McGraw-Hill, New York, 1986, p. 84

10 'The Wages of Zealotry', *The New York Times*, February 20, 1983, http://www.nytimes.com/1983/02/20/opinion/the-wages-of-zealotry.html Accessed November 29, 2016

11 Barry Boyer & Errol Meidinger, 'Privatizing Regulatory Enforcement: A Preliminary Assessment of Citizen Suits Under Federal Environmental Laws', *Buffalo Law Review*, vol. 34 (1985), pp. 863–964, http://web2.law.buffalo.edu/faculty/meidinger/scholarship/citsuit.pdf Accessed March 16, 2015

12 Jeffrey G. Miller, *Citizen Suits: Private Enforcement of Federal Pollution Control Laws*, Wiley, New York, 1987

13 Paul Clancy, 'Hot Times in Ham Town' *Chesapeake Bay Magazine*, March 2008, http://www.chesapeakeboating.net/Publications/Chesapeake-Bay-Magazine/1999/From-the-Chesapeake-Bay-Magazine-Archives/Destination-Smithfield-VA.aspx Accessed November 29, 2016 — a generally positive piece with good detail on the historic town

14 Jeff Tietz, 'Boss Hog: The Dark Side of America's Top Pork Producer', *Rolling Stone*, December 14, 2006, http://www.rollingstone.com/culture/news/boss-hog-the-dark-side-of-americas-top-pork-producer-20061214 Accessed March 19, 2015. Smithfield issued a document on their website (now removed) to counter points in the Tietz article, and this pathogen level was one of the facts they countered, directing you to the North Carolina University website, where they have funded research, for a counter argument. The clearest such is M.D. Sobsey et al., 'Pathogens in Animal Waste and the Impacts of Waste Management Practices on Their Survival, Transport, and Fate', http://www.cals.ncsu.edu/waste_mgt/natlcenter/whitepapersummaries/pathogens.pdf, which suggests 'high concentrations (millions to billions per gram of wet weight feces) of human pathogens (disease-causing microorganisms)'. This is in fact more startling than the figure Smithfield challenges. Also startling are the number of Tietz's grievous complaints they don't even try to challenge

15 Tietz, op. cit.

16 Miller, op. cit., p. 12, figures based on a study by the Environmental Law Institute

17 ibid.

18 Carol Lichti, 'The Man Behind a $50 Million Naming-rights Offer and a $6 Billion Company', *Inside Business: The Hampton Roads Business Journal*, February 4, 2002, http://insidebiz.com/news/man-behind-50-million-naming-rights-offer-and-6-billion-company Accessed March 26, 2015

19 Chesapeake Bay Foundation v. Gwaltney of Smithfield, 611 F. Supp. 1542 (ED Va. 1985), http://law.justia.com/cases/federal/district-courts/FSupp/611/1542/2003260/ Accessed April 1, 2015

20 Diana L. Lee, 'Gwaltney of Smithfield, Ltd. v. Chesapeake Bay Foundation: Its Implications for Citizen Suits under the Clean Water Act', *Ecology Law Quarterly*, vol. 16, no. 2 (March 1989), p. 571, http://scholarship.law.berkeley.edu/cgi/viewcontent.cgi?article=1358&context=elq Accessed March 30, 2015

21 Michael J. Flannery, 'The Virginia Environmental Endowment: Past and Future', *William & Mary Environmental Law & Policy Review*, vol. 14, no. 1 (1989), pp. 37–50, http://scholarship.law.wm.edu/cgi/viewcontent.cgi?article=1392&context=wmelpr Accessed April 1, 2015

22 Chesapeake Bay Foundation v. Gwaltney of Smithfield, op. cit.

23 118 Cong. Rec. 33,700, reprinted in *A Legislative History of the Water Pollution Control Act Amendments of 1972*, vol. 1, US Government Printing Office, Washington, 1973, p. 179

24 Lee, op. cit.

25 Case summary testimony recorded at https://casetext.com/case/chesapeake-bay-v-gwaltney-of-smithfield Accessed March 27, 2015

26 Tietz, op. cit.

27 Boyer & Meidinger, op. cit.

28 ibid.

29 Churchville, op. cit.

30 ibid.

31 Associated Press, 'Steel Company Sued for Polluting Bay', *Observer-Reporter*, Washington, PA, February 16, 1987

32 Churchville, op. cit.

33 Associated Press, op. cit.

34 Chesapeake Bay Foundation v. Bethlehem Steel Corp., 652 F. Supp. 620 (D. Md 1987), January 30, 1987, http://law.justia.com/cases/federal/district-courts/FSupp/652/620/2305538/ Accessed November 29, 2016

35 Marcia R. Gelpe & Janis L. Barnes, 'Penalties in Settlements of Citizen Suit Enforcement

Actions Under the Clean Water Act', *William Mitchell Law Review*, vol. 16 (1990), pp. 1025–1040, http://papers.ssrn.com/sol3/papers.cfm?abstract_id=1703117 Accessed November 29, 2016

36 Graham Thompson & Jon Turk, *Earth Science and the Environment*, Cengage Learning, 2006 p. 415

37 Miller, op. cit., pp. 12–13

38 Michael Gorn, *William D. Ruckelshaus: Oral History Interview*, Environmental Protection Agency, 1993, http://www2.epa.gov/aboutepa/william-d-ruckelshaus-oral-history-interview Accessed March 27, 2015

39 Boyer & Meidinger, op. cit.

40 Robert Walters, 'Groups Take Action Against Anti-pollution Law Violators', *The Baytown Sun*, Texas, April 6, 1987

41 Stuart Taylor, Jr, 'Supreme Court Roundup; Citizens' Suits in Pollution Cases Are Limited', *The New York Times*, December 2, 1987, http://www.nytimes.com/1987/12/02/us/supreme-court-roundup-citizens-suits-in-pollution-cases-are-limited.html Accessed April 2, 2015

42 Churchville, op. cit.

43 Tietz, op. cit.

44 Lichti, op. cit.

45 'State of the Bay' reports on the Chesapeake Bay Foundation website: http://www.cbf.org/news-media/newsroom/cbf-reports

2. Where the Wild Things Are

1   For a recent and detailed survey of the Park, see N. Anderson, 'Sycamore Canyon Natural History: A Report to the Riverside Municipal Museum', August 10, 2015, http://www.krazykioti.com/articles/sycamore-canyon-natural-history/ Accessed March 30, 2016

2   George Johnson, 'The Jaguar and the Fox', *The Atlantic Monthly*, vol. 287, no. 1 (July 2000), pp. 82–85

3   James Thornton, *A Field Guide to the Soul*, Bell Tower, New York, 1999, p. 100

4   Daniel Rubinoff, in his 'Evaluating the California Gnatcatcher as an Umbrella Species for Conservation of Southern California Coastal Sage Scrub', *Conservation Biology*, vol. 15, no. 5 (October 2001), pp. 1374–1383, noted how the US endangered species list contained '331 vertebrates but only 45 insects and arachnids'. His study tested the success of the California gnatcatcher as an 'umbrella species' by seeing how well three invertebrates were preserved in the region. Of the 50 patches he surveyed, he found the gnatcatcher in 48 of them, and his invertebrates in considerably fewer. The larger the patch, the more comprehensive were the numbers of endangered creatures inside it. A

lawyer builds a narrative, and the brutal truth is that an appealing bird or mammal adds an emotional punch that even a rat or a lizard lacks. The chances of halting multi-million dollar developments to preserve an arachnid are sadly minimal

5 Jonathan L. Atwood, 'Speciation and Geographic Variation in Black-tailed Gnatcatchers', *Ornithological Monographs*, no. 42 (1998). See also Atwood, 'Subspecies Limits and Geographic Patterns of Morphological Variation in California Gnatcatchers (Polioptila californica)', *Bulletin of the Southern California Academy of Sciences*, vol. 90 (1991), no. 3, pp. 118–133. Atwood's work supported the 1993 listing of the bird as an endangered species

6 Jeffery V. Wells, *Birder's Conservation Handbook: 100 North American Birds at Risk*, Princeton University Press, 2010, p. 288

7 Rita Beamish (Associated Press), 'Feds OK Harming Tiny California Bird in Order to Help It', *Ukiah Daily Journal*, March 26, 1993, p. 3

8 For a discussion of this legal challenge, see Brigid McCormack, 'Developers Claim Bird Species Doesn't Even Exist: Guest Commentary', *Los Angeles Daily News*, February 20, 2015, http://www.dailynews.com/opinion/20150220/developers-claim-bird-species-doesnt-even-exist-guest-commentary Accessed July 23, 2015

9 Press Release, Centre for Biological Diversity, 'Review Initiated Over Protected Status of Coastal California Gnatcatcher', December 30, 2014, http://www.biologicaldiversity.org/news/press_releases/2014/california-gnatcatcher-12-30-2014.html Accessed July 23, 2015

3. Leveraging Alaska

1 Mark Levinson, *The Great A&P and the Struggle for Small Business in America*, New York, Hill & Wang, 2011

2 For a clear ecologist's view of Alaska, see Carl Safina, *The View from Lazy Point*, Henry Holt, New York, pp. 154–182

4. The Lie of the Land

1 Interview with John Adams, Alaska, August 2015

2 Tim Smedley, 'At Work with the FT Interview: James Thornton, ClientEarth', *Financial Times*, May 12, 2016

3 James Thornton, *A Field Guide to the Soul*, Bell Tower, New York, 2000, p. 178

4 Mitchell Thomashow, *Ecological Identity: Becoming a Reflective Environmentalist*, Cambridge, MA, MIT Press, 1996, p. 148

5 Ludwig Krämer, *EU Environmental Law*, 7th ed., Sweet & Maxwell, London, 2011, p. v

6 Stefan Scheuer, ed., *EU Environmental Policy Handbook: A Critical Evaluation of EU Environmental Legislation, European Environmental Bureau*, European Environmental

Bureau, Utrecht, 2005, p. 9, http://www.wecf.eu/cms/download/2004-2005/EEB_Book. pdf Accessed May 16, 2016

7 John E. Bonine, 'Public Interest Lawyers — Global Examples and Personal Reflections', *Widener Law Review*, vol 10 (2004), no. 2, p. 451

8 http://en.frankbold.org/about-us/what-we-do Accessed May 16, 2016

9 Interview with Jan Šrytr, Kraków, July 2014

10 Interview with Kristina Sabova, Kraków, July 2014

11 Yves Dezalay, 'From a Symbolic Boom to a Marketing Bust: Genesis and Reconstruction of a Field of Legal and Political Expertise at he Crossroads of a Europe Openign to the Atlantic', *Law & Social Inquiry*, vol. 32, no. 1 (Winter 2007), pp. 161–181

12 Robert Gottlieb, *Forcing the Spring: The Transformation of the American Environmental Movement*, Island Press, Washington, DC, 1993, p. 142

13 David Coen, 'Lobbying in the European Union: Briefing paper', European Parliament, November 2007, http://www.europarl.europa.eu/RegData/etudes/etudes/join/2007/393266/IPOL-AFCO_ET(2007)393266_EN.pdf Accessed September 4, 2016

14 Nicolas de Sadeleer et al., *Access to Justice in Environmental Matters and the Role of NGOs: Empirical Findings and Legal Appraisal*, UNECE, Groningen, 2005, pp. 14–15, http://www.unece.org/fileadmin/DAM/env/pp/compliance/C2008-23/Amicus%20brief/AnnexHSadeleerReport.pdf Accessed May 16, 2016

15 Rachael Smith et al., 'Personality Trait Differences between Traditional and Social Entrepreneurs', *Social Enterprise Journal*, vol. 10, no. 3, pp. 200–221, http://doi.org/10.1108/SEJ-08-2013-0033 Accessed April 2, 2016

16 Robert A. Baron, 'Opportunity Recognition as Pattern Recognition: How Entrepreneurs "Connect the Dots" to Identify New Business Opportunities', *Academy of Management Perspectives*, vol. 20, no. 1 (February 2006), pp. 104-111

17 ibid., as quoted by Robert A. Baron

18 Chris Hilson, 'New Social Movements: The Role of Legal Opportunity', *Journal of European Public Policy*, vol. 9 (2002), no. 2, pp. 238–255. See also: Lisa Vanhala, 'Legal Opportunity Structures and the Paradox of Legal Mobilization by the Environmental Movement in the UK', *Law & Society Review*, vol. 46, no. 3 (September 2012), pp. 523–556. Lisa Vanhala, a scholar who looks at the relationship between NGOs, law, and social change, observed that Greenpeace was 'the first organization to work closely with lawyers and establish in-house legal advice. The reason cited for this was Greenpeace's use of civil disobedience tactics and related encounters with criminal law.' The group was also one of the first to bring cases, seeking a judicial review in 1993, while Friends of the Earth was seen 'as being law-savvy from the late 1980s onwards, primarily because of the long-standing presence of lawyers on its board of trustees'. She notes how ClientEarth, 'a

public interest law-devoted organization' resembling the 'American-style' legal NGOs that have thus far been unknown in the UK ... suggests that changes are occurring in the UK environmental movement'. While Greenpeace and Friends of the Earth did bring cases in the UK to advance their objectives, sometimes achieving landmark results, neither group made law central to the planning and realisation of all activities of the group, from campaigning, to seeking changes in laws, or taking steps to ensure that the laws relevant to an area of campaigning were implemented and enforced

19 The Attorney General Sir Hartley Shawcross, Hansard, HC Deb 15 December 1948 vol. 459 cc1221–327, http://hansard.millbanksystems.com/commons/1948/dec/15/legal-aid-and-advice-bill Accessed March 12, 2015

20 Dezalay, op. cit., p. 177

21 Tate Williams, 'Where the Hell is all the Climate Funding', *Inside Philanthropy*, April 22, 2015, http://www.insidephilanthropy.com/home/2015/4/22/where-the-hell-is-all-the-climate-funding.html Accessed September 2, 2016

22 From an interview with Phil Michaels, London, 11 March 2014. Very sadly, Phil died the following Autumn

23 Interview with John Adams, Juneau, August 2015

24 Zoe Williams, 'Alastair Campbell', *The Guardian*, February 28, 2015, http://www.theguardian.com/politics/2015/feb/27/alastair-campbell-interview Accessed March 3, 2015

25 United Nations Treaty Collection, https://treaties.un.org/Pages/ViewDetails.aspx?src=IND&mtdsg_no=XXVII-13&chapter=27&clang=_en Accessed March 13, 2015

26 Carol Day, 'Is the Aarhus Costs Regime EU Law Compliant?', March 5, 2014, http://www.leighday.co.uk/News/2014/March-2014/Is-the-Aarhus-costs-regime-EU-law-compliant- Accessed March 13, 2015

5. An Air That Kills

1 Heather Walton et al., 'King's College London Report on Mortality Burden of NO_2 and $PM_{2.5}$ in London', Kings College London, for Transport for London and the Greater London Authority, July 14, 2015, p. 39, http://www.scribd.com/doc/271641490/King-s-College-London-report-on-mortality-burden-of-NO2-and-PM2-5-in-London Accessed October 4, 2015

2 Timothy O'Riordan, 'Public Interest Environmental Groups in the United States and Britain', *Journal of American Studies*, vol. 13, no. 3 (December 1979), pp. 409–438

3 OSPAR's reply was one of 74 additional documents reviewed at the United Nations, where Bob Latimer's battle to protect fish was viewed as a classic case study. See the ClientEarth letter to Jeremy Wates: http://www.unece.org/

fileadmin/DAM/env/pp/compliance/C2008-33/response/From%20communicant/
FrCommReC3320090609ClientEarthresponseQuestions.pdf Accessed June 2, 2015

4 The UK alliance that had been working in the Aarhus arena for some years, known as
CAJE, chose to hold back. Phil Michaels, formerly head of legal strategy at Friends of the
Earth and a founder member of this group, explained to me how they preferred to keep
to their own strategy, which had been continuous since 1995. The case was submitted in
December 2008, and CAJE joined as an amicus in May 2009

5 Press Release, ClientEarth, 'Environmental Justice Cases "Will Establish Fundamental
Rights for Green Groups"', September 18, 2009, http://www.clientearth.org/reports/
Press%20release%20-%20ClientEarth%20Aarhus%20Compliance.pdf Accessed June
2, 2015

6 Blackstone Chambers, 'James Eadie QC', https://www.blackstonechambers.com/
barristers/james-eadie-qc/ Accessed June 11, 2015

7 Aarhus Convention Compliance Committee, 'Report of the Compliance Committee
on Its 29th Meeting, Addendum, Findings and Recommendations with Regard to
Communication ACCC/C/2008/33 Concerning Compliance by the United Kingdom
of Great Britain and Northern Ireland', ECE/MP.PP/C.1/2010/6/Add.3, August 24,
2011, p. 9, http://www.unece.org/fileadmin/DAM/env/pp/compliance/C2008-33/
Findings/ece_mp.pp_c.1_2010_6_add.3_eng.pdf Accessed June 11, 2015

8 A. Andrusevych et al. (eds), *Case Law of the Aarhus Convention Compliance Committee
(2004–2011)*, 2nd ed., Academic Network on the European Social Charter and Social
Rights, Lviv, Ukraine, p. 172

9 'David v. Goliath Win over Toxic Waste', *The Shields Gazette*, September 21, 2010,
http://www.shieldsgazette.com/news/local-news/david-v-goliath-win-over-toxic-
waste-1-2038768 Accessed June 2, 2015

10 Interview with Jan Šrytr, Kraków, July 2014

11 A UK government letter of January 28, 1994, included in documents obtained by Clean
Air in London under the Freedom of Information Act, found at http://cleanair.london/
wp-content/uploads/CAL-306-DoH-FoI-reply-re-diesel-vs-climate-200715_Annex-A.pdf
Accessed April 12, 2016

12 John Vidal, 'All Choked Up: Did Britain's Dirty Air Make Me Dangerously Ill?', *The
Guardian*, June 20, 2015, http://www.theguardian.com/global/2015/jun/20/britain-
london-pollution-air-quality-health Accessed April 13, 2016

13 Air Quality Expert Group, *Nitrogen Dioxide in the United Kingdom*, Defra, 2004, http://
uk-air.defra.gov.uk/assets/documents/reports/aqeg/nd-summary.pdf Accessed June 18,
2015

14 Richard Howard et al., 'Up in the Air: How to Solve London's Air Quality Crisis: Part
2', Kings College London, Policy Exchange, and Capital City Foundation, 2016, p. 6

https://policyexchange.org.uk/wp-content/uploads/2016/09/up-in-the-air-part-2.pdf
Accessed April 8, 2016

15 Frank Kelly, 'Gasping for Air', *The Sunday Times,* May 28, 2015, p. 4

16 Press Release, British Thoracic Society, 'Exposure to Diesel Pollution on Oxford Street Presents Health Risks to People with Lung Problems and Healthy Pedestrians', December 5, 2014, https://www.brit-thoracic.org.uk/document-library/news/press-releases-2014-wm/diesel-pollution-on-oxford-street/ Accessed June 29, 2015

17 Jonathan Leake, 'Oxford Street Worst in the World for Diesel Pollution', *The Sunday Times*, July 6, 2014, from an interview with Kings College scientist David Carslaw

18 Jonathan Leake, 'Dirty Diesel Death Toll Hits 60,000', *The Sunday Times*, November 30, 2014

19 Nina Chestney, 'Parts of UK Will Not Meet EU Pollution Limits until after 2030', Reuters, July 10, 2014, http://www.trust.org/item/20140710123740-u1efp?view=print Accessed June 24, 2015

20 Matt McGrath, 'UK Admits That Air Quality Targets Will Be Missed by 20 Years', BBC News, July 10, 2014, http://www.bbc.co.uk/news/science-environment-28255246 Accessed June 28, 2015

21 Court of Justice of the European Union, Press Release no. 153/14, Luxembourg, November 19, 2014

22 http://bankwatch.org/documents/KrakowSmogAlert-timeline.pdf Accessed November 30, 2016

23 The Supreme Court, 'Press Summary', April 29, 2015, https://www.supremecourt.uk/decided-cases/docs/UKSC_2012_0179_PressSummary.pdf Accessed July 2, 2015

24 James Thornton, 'Can We Catch Up? How the UK Is Falling Behind on Environmental Law', *Environmental Law & Management,* vol. 27 (2015), no. 5, pp. 193–199

25 Jonathan Leake, 'Cleaner Air in Cities "Blocked by Treasury"', *The Sunday Times*, October 23, 2016

26 Damian Carrington, 'Diesel Vehicles Face Charges after UK Government Loses Air Pollution Case', *The Guardian*, November 2, 2016

27 Comprising 467,000 from $PM_{2.5}$ alone, plus 71,000 who died early from NO_2 pollution, in 2013, and 'These figures do not show significant changes over the years.' European Environment Agency, 'Air Quality in Europe — 2016 Report', Luxembourg, 2016, http://www.eea.europa.eu/publications/air-quality-in-europe-2016 Accessed December 7, 2016

28 ibid., pp. 11–12

29 Chris Grayling 'The Airports Commission Report', letter of November 8, 2016,

http://www.parliament.uk/documents/commons-committees/environmental-audit/correspondence/161108-Chris-Grayling-to-Mary-Creagh-Airports-Commission-Report.pdf Accessed November 26, 2016

30 Nicholas Cecil, 'Government Backed Heathrow Airport Third Runway "Using Old Pollution Data"', *Evening Standard*, November 25, 2016, http://www.standard.co.uk/news/transport/government-backed-heathrow-third-runway-using-old-pollution-data-a3405001.html Accessed November 26, 2016

31 European Environment Agency, op. cit., pp. 11, 60 for the figures

The sheriff comes to town (James Thornton)

1 World Health Organization, 'Almost 600,000 Deaths Due to Air Pollution in Europe: New WHO Global Report', March 25, 2014, http://www.euro.who.int/en/health-topics/environment-and-health/air-quality/news/news/2014/03/almost-600-000-deaths-due-to-air-pollution-in-europe-new-who-global-report Accessed April 3, 2016

6. Leaving Plenty More Fish in the Sea

1 Boris Worm et al., 'Impacts of Biodiversity Loss on Ocean Ecosystem Service', *Science*, vol. 315, no. 5800 (November 3, 2006), pp. 787–790

2 Erik Stokstad, 'Global Loss of Biodiversity Harming Ocean Bounty', *Science*, vol. 315, no. 5800 (November 3, 2006), p. 745

3 Food and Agricultural Organization of the UN, *The State of World Fisheries and Aquaculture 2006*, Rome, 2007, p. 33

4 The other five strategic points: Is it possible to get any leverage on that problem with limited resources, and so form an action plan that will succeed? Is there a forum in which you can operate successfully? Is there a high probability of success? Has anybody else pre-empted the work? Is there funding?

5 George Perkins Marsh, *Man and Nature*, University of Washington, Seattle, 2003 [1864], p. 101

6 David Lowenthal, 'Introduction to the 2003 Edition', in Marsh, op. cit.

7 Will Steffen et al., 'The Anthropocene: Conceptual and Historical Perspectives', *Philosophical Transactions of the Royal Society A*, vol. 369, no. 1938 (March 13, 2011), pp. 842–886

8 Paul J. Crutzen & Eugene F. Stoermer, 'The Anthropocene', IGBP Newsletter, 2000, included in Libby Robin et al. (eds), *The Future of Nature*, Yale University Press, New Haven, 2013, p. 485

9 Gilbert St-Pierre, *Spawning Locations and Season for Pacific Halibut*, Scientific Report No. 70, International Pacific Halibut Commission, Seattle, 1984

10 Cathy Roheim Wessells & James L. Anderson, 'Innovations and Progress in Seafood Demand and Market Analysis', *Marine Resource Economics*, vol. 7, no. 4 (Winter 1992), pp. 209–228

11 Robert N. Stavins, 'The Problem of the Commons: Still Unsettled after 100 Years', *American Economic Review*, vol. 101, no. 1 (February 2011), pp. 81–108

12 Sharon Levy, 'Catch Shares Management', *BioScience*, vol. 60, no. 10 (November 2010), pp. 780–785

13 Olivier Thébaud et al., 'From Anecdotes to Scientific Evidence? A Review of Recent Literature on Catch Share Systems in Marine Fisheries', *Frontiers in Ecology & the Environment*, vol. 10, no. 8 (October 2012), pp. 433–437

14 Levy, op. cit.

15 M. Aaron MacNeil et al., 'Transitional States in Marine Fisheries', *Philosophical Transactions of the Royal Society B*, vol. 365 (2010), no. 1558, pp. 3753–3763 (p. 3757)

16 J. Samuel Barkin & Elizabeth R. DeSombre, *Saving Global Fisheries*, MIT Press, Cambridge, MA, 2013, pp. 62, 65

17 Ocean2012, http://www.pewtrusts.org/en/archived-projects/ocean2012/about Accessed February 16, 2015

18 Adam Soliman, 'Individual Transferable Quotas in World Fisheries: Addressing Legal and Rights-based Issues', *Ocean & Coastal Management*, vol. 87 (January 2014), pp. 102–113

19 'Cool It: Cleaning Up the Old Act,' *The Economist*, August 31, 1991, http://www.edf.org/sites/default/files/11682_The_Economist_Cool_It.pdf Accessed February 11, 2015

20 'Reforming European Fisheries', http://www.edf.org/oceans/reforming-european-fisheries-sharing-knowledge-and-expertise-ground Accessed February 11, 2015

21 Christopher Costello et al., 'Can Catch Shares Prevent Fisheries Collapse?' *Science*, vol. 321, no. 5896 (September 19, 2008), pp. 678–681

22 Levy, op. cit., p. 785

23 Soliman, op. cit.

24 Gina Hanrahan, 'Reform of the CFP — Council Reaches Agreement, but Is It Green Enough?', The Institute of International and European Affairs, July 12, 2012, http://www.iiea.com/blogosphere/reform-of-the-cfp--council-reaches-agreement-but-is-it-green-enough Accessed June 16, 2015

25 Rob Wilson, 'Richard Benyon MP on Having "the Crap" Beaten Out of Him', April 2, 2013, http://www.totalpolitics.com/articles/interview/richard-benyon-mp-having-crap-beaten-out-him Accessed November 30, 2016

26 Demersal fish live on or near the seabed, and are often caught together

27 William Robinson, 'Drafting European Legislation in the European Commission: A Collaborative Process', *The Theory and Practice of Legislation*, vol. 2 (2014), no. 3, pp. 249–272

28 Paul Greenberg & Boris Worm, 'When Humans Declared War on Fish,' *The New York Times*, May 18, 2015, http://www.nytimes.com/2015/05/10/opinion/sunday/when-humans-declared-war-on-fish.html Accessed April 27, 2016

29 Christopher Costello et al., 'Global Fishery Prospects under Contrasting Management Regimes', *Proceedings of the National Academy of Sciences*, vol. 113, no. 18, published online March 28, 2016, http://www.pnas.org/content/early/2016/03/29/1520420113. full Accessed April 27, 2016

30 Bob Murphy, 'Fisheries Study Shows 2/3 of Fish Stocks Declining, Researcher Says', CBC News, April 5, 2016, http://www.cbc.ca/news/canada/nova-scotia/boris-worm-study-fisheries-management-1.3522301 Accessed April 27, 2016

Setting up in Brussels (James Thornton)

1 The two cases in the Europe's highest court, the European Court of Justice, are: ClientEarth v. European Commission, Case C-612/13 P (2015) (Commission required to release studies showing whether EU countries had properly transposed EU law into the domestic law); and ClientEarth v. European Food Safety Authority (EFSA), Case C-615/13 P (2015) (EFSA required to disclose the input of experts in its rule-making, when those experts work for industries regulated by EFSA rules). Both of these cases were argued pro bono by Pierre Kirch, partner in the Paris office of Paul Hastings

2 The file for the proceedings of the Aarhus Convention Compliance Committee in the complaint that ClientEarth made against the EU is referred to as ACCC/C/2008/32. All of the proceedings including the decision against the EU may be found at http://www.unece.org/env/pp/compliance/Compliancecommittee/32TableEC.html Accessed November 30, 2016

3 Laurens Ankersmit & Karla Hill, 'Legality of Investor State Dispute Resolution Mechanism under EU Law', ClientEarth, October 22, 2015, http://www.documents.clientearth.org/library/download-info/legality-of-investor-state-dispute-settlement-under-eu-law/ Accessed September 7, 2016; Laurens Ankersmit, 'The Compatability of Investment Arbitration in EU Trade Agreements with the EU Judicial System', *Journal for European Environmental & Planning Law*, vol. 13 (2016), no. 1, pp. 46–63

4 Jennifer Rankin, 'Compromise Opens Way to Signing of EU Trade Agreement with Canada', *The Guardian,* October 28, 2016, p. 29

7. Coals of Fire

1 Mike Schwarz, 'Why Did Ratcliffe Defence Fail where Kingsnorth Six Succeeded?', *The Guardian*, December 16, 2010, http://www.theguardian.com/environment/cif-green/2010/dec/16/ratcliffe-trial Accessed December 1, 2015

2   James Hansen, 'Twenty Years Later: Tipping Points Near on Global Warming', *The Huffington Post*, July 1, 2008, http://www.huffingtonpost.com/dr-james-hansen/twenty-years-later-tippin_b_108766.html Accessed July 27 2015

3   Dieter Helm, *The Carbon Crunch*, Yale University Press, New Haven, 2012, p. 34

4   James Hansen, 'Statement of Witness James E. Hansen to the Kingsnorth Trial', http://www.greenpeace.org.uk/files/pdfs/climate/hansen.pdf Accessed October 24, 2014

5   Interview with Matt Phillips, April 2014

6   'Tim Malloch', Mishcon de Reya, http://www.mishcon.com/people/tim_malloch Accessed November 30, 2016

7   'Poland — Energy Mix Fact Sheet', European Commission, January 2007, http://ec.europa.eu/energy/energy_policy/doc/factsheets/mix/mix_pl_en.pdf Accessed October 1, 14

8   Donald Trusk, 'A United Europe Can End Russia's Energy Stranglehold', *Financial Times*, April 21, 2014, https://www.ft.com/content/91508464-c661-11e3-ba0e-00144feabdc0 Accessed November 17, 2016

9   'In with the New', *The Economist*, June 28, 2014, http://www.economist.com/news/special-report/21604686-traditional-industries-are-declining-outsourcing-offshoring-and-subcontracting-are Accessed October 1, 14

10 Reuters, 'Poland Proposes Plan to Provide Cash for Coal Miner Kompania Weglowa', September 7, 2015, http://www.reuters.com/article/2015/09/07/poland-mining-idUSL5N11D3S620150907 Accessed October 18, 2015

11 Adam Easton, 'Polish State Energy Companies to Invest $624 Mil in Coal Miner Kompania Weglowa', S&P Global Platts, April 26, 2016, http://www.platts.com/latest-news/coal/warsaw/polish-state-energy-companies-to-invest-624-mil-26427960 Accessed December 3, 2016

12 Marcin Stoczkiewicz & Ilona Jedrasik, 'Understanding the Polish Anti-climate Crusade', *Energy Post*, October 23, 2014, http://www.energypost.eu/understanding-polish-anti-climate-crusade-2/ Accessed October 28, 2014

13 Helm, op. cit., pp. 34–35

14 Paul Baruya, 'Losses in the Coal Supply Chain', IEA Clean Coal Centre Profiles, no. 12/18, December 2012

15 'Jan Kulczyk', *Forbes*, http://www.forbes.com/profile/jan-kulczyk/ Accessed March 29, 2016

16 C.S.A. van Koppen & Wiliam T. Markham (eds), *Protecting Nature: Organizations and Networks in Europe and the USA*, Edward Elgar, Cheltenham, 2008

17 Agnieszka Barteczko, 'Polish Green Campaigners in Court Win over Coal Plant', Reuters, February 14, 2013, http://www.reuters.com/article/poland-kulczyk-idUSL5N0BE9FU20130214 Accessed March 30, 2016

18 Diarmaid Williams, 'Another Setback for Giant Polish Coal-fired Power Plant', *Power Engineering International*, February 6, 2014, http://www.powerengineeringint.com/articles/2014/02/another-setback-for-giant-polish-coal-fired-power-plant.html Accessed March 30, 2016

19 'Polnoc Power Station', SourceWatch, January 2016, http://www.sourcewatch.org/index.php?title=Polnoc_Power_Station&oldid=689768 Accessed March 30, 2016

20 European Environment Agency, 'Air Quality in Europe — 2016 Report', Luxembourg, 2016, http://www.eea.europa.eu/publications/air-quality-in-europe-2016 Accessed December 7, 2016

21 ClientEarth, 'Europe's Largest New Coal Power Plant Stopped', December 6, 2017, http://www.clientearth.org/europes-largest-new-coal-power-plant-stopped/ Accessed December 7, 2016

22 'Poles Going Green', Radio Poland, November 20, 2015, http://www.thenews.pl/1/9/Artykul/229659,Poles-going-green Accessed December 9, 2015

8. The Forests of Africa

1 Manesseh Azure Awuni, 'Rescuing Achimota Forest Reserve', *Graphic Online*, November 18, 2013, http://graphic.com.gh/features/opinion/4939-rescuing-achimota-forest-reserve.html Accessed September 7, 2015

2 Kofi Oteng Kufuor, 'Forest Management in Ghana: Towards a Sustainable Approach', *Journal of African Law*, vol. 44, no. 1 (2000), pp. 52–64

3 W.G. Sebald, *The Rings of Saturn*, Vintage, London, 2002, p. 122

4 ClientEarth, *An Overview of the Legal Framework of the Forest and Wildlife Sector*, London, November 2013, p. 4

5 Bongo District Assembly, 'Bongo District: Environmental Situation', 2006, http://bongo.ghanadistricts.gov.gh/?arrow=atd&_=103&sa=4956 Accessed September 16, 2015

6 ClientEarth, 'Resources — Ghana', http://www.clientearth.org/ghana/#resources Accessed November 30, 2016

7 Jens Friis Lund et al., 'The Political Economy of Timber Governance in Ghana', *ETFRN News*, no. 53 (April 2012), pp. 117–126

8 Judith Carney & Marlène Elias, 'Revealing Gendered Landscapes: Indigenous Female Knowledge and Agroforestry of African Shea', *Canadian Journal of African Studies*, vol. 40 (2006), no. 2, pp. 235–267

9 Bernice Agyekwena, 'The Sheanut Tree, the Wonder Tree', Bernice Agyekwena's Blog, January 4, 2011, https://berniceagyekwena.wordpress.com/2011/01/04/the-sheanut-tree-the-wonder-tree/ Accessed September 25, 2015

10 Clara Melot, 'Protecting the Shea Tree', ClientEarth, July 18, 2014, http://www.blog.

clientearth.org/shea-problematic/ Accessed September 25, 2015

11 Fred Pearce, 'In Ghana's Forests, Should Chainsaw Loggers Be Legalized?', *Yale Environment 360*, August 16, 2012, http://e360.yale.edu/feature/in_ghanas_forests_should_chainsaw_loggers_be_legalized/2562/ Accessed October 1, 2015

12 Lund, op. cit.

13 Daron Acemoglu & James A. Robinson, *Why Nations Fail: The Origins of Power, Prosperity, and Poverty*, Profile, London, 2013, p. 76

14 Jared Diamond, 'What Makes Countries Rich or Poor?', *The New York Review of Books*, June 7, 2012, http://www.nybooks.com/articles/archives/2012/jun/07/what-makes-countries-rich-or-poor/ Accessed October 2, 2015

15 Food and Agricultural Organization of the UN, *Global Forest Resources Assessment 2015: How Are the World's Forests Changing?*, Rome, 2015, http://www.fao.org/3/a-i4793e.pdf Accessed October 2, 2015

16 Damian Carrington, 'Indonesia Cracks Down on Deforestation in Symbolic U-turn', *The Guardian*, November 27, 2014

Making laws work: implementation (James Thornton)

1 Elizabeth Hiester et al., 'REACH Registration and Endocrine Disrupting Chemicals', ClientEarth, July 2013, http://www.clientearth.org/reports/reach-registration-and-endocrine-disrupting-chemicals.pdf Accessed April 14, 2016

2 ClientEarth v. European Food Safety Authority (EFSA), Case C-615/13 P (2015)

9. The Dragon Awakes

1 Edward Wong, 'China Approves 155 New Coal Power Plants', *The New York Times*, November 11, 2015

2 Henry Kissinger, *On China*, Penguin, London, 2012, p. 16

3 ibid., p. 2

4 For an account of the Chinese scholar Yan Jisheng's archival research to restore the factual record, see Tania Branigan, 'China's Great Famine: The True Story', *The Guardian*, January 1, 2013

5 Kissinger, op. cit., pp. 15–16

6 Richard Wike & Bridget Parker, 'Corruption, Pollution, Inequality Are Top Concerns in China,' *Global Attitudes and Trends*, Pew Research Centre, September 23, 2015 http://www.pewglobal.org/2015/09/24/corruption-pollution-inequality-are-top-concerns-in-china/ Accessed December 24, 2015

7 Ying Zhu & Bruce Robinson, 'Critical Masses, Commerce, and Shifting State-Society

Relations in China', *The China Beat*, February 17, 2010, http://www.thechinabeat. org/?p=1526 Accessed December 24, 2015

8 Charles Clover, 'China Web Tsar Admits Censorship Troubles', *Financial Times*, December 9, 2015

9 CCICED, 'Summary Record', *The China Council for International Cooperation on Environment and Development: The Third Meeting of the Fifth Phase*, Beijing, 2015

10 For the Chinese statistics, see Chaohua Wang, 'I'm a Petitioner — Open Fire!', *London Review of Books*, November 5, 2015, p. 13

11 Jiahua Pan, *China's Environmental Governing and Ecological Civilization*, Springer, Berlin, 2016, p. 40

12 Pan, op. cit., p. 35

13 Laozi/Lao-Tzu, trs. James Legge, *The Tao Teh King, or The Tao and Its Characteristics*, The Tao Info, http://www.thetao.info/english/english.htm

14 Pan, op. cit., p. 39

15 Sam Geall, 'Interpreting Ecological Civilisation (Part One)', chinadialogue, July 6, 2015, https://www.chinadialogue.net/article/show/single/en/8018-Interpreting-ecological-civilisation-part-one- Accessed February 25, 2016

16 Xi Jinping, 'Talks on Ecological Civilization', August 29, 2014, ref. & trs. Zhihe Wang et al., 'The Ecological Civilization Debate in China', *Monthly Review*, vol. 66, no. 6 (November 2014), http://monthlyreview.org/2014/11/01/the-ecological-civilization-debate-in-china/ Accessed February 25, 2016

17 Editorial, 'Rules of the Party', *The Economist*, November 1, 2014, http://www.economist. com/news/china/21629528-call-revive-countrys-constitution-will-not-necessarily-establish-rule-law-rules Accessed February 25, 2016

18 Wang et al., op. cit.

19 Wang Canfa, 'The Current Situation and Future Anticipation of China's Modern Environmental Legislation', March 11, 2011, ref. & trs. Wang et al., op. cit.

20 Tseming Yang, 'Getting an Army Ready to Fight the War', *The Environmental Forum*, vol. 32, no. 5 (September/October 2015)

21 Dai Jie, 'The Influence of the World on China's Environmental Laws and the Correction Way', *Journal of Xihua Normal University*, vol. 1 (2013), ref. & trs. Wang et al.

22 Editorial, op. cit.

23 Opinion Line, 'Everyone Responsible for Creating a Healthy Ecological Civilization', *China Daily*, 7 May 2015, http://www.chinadaily.com.cn/opinion/2015-05/07/content_20642952.htm Accessed February 25, 2016

24 Editorial, 'Bo Xilai Gets Life in Prison', *China Daily USA*, September 23, 2013, http://usa.chinadaily.com.cn/china/2013-09/22/content_16984347.htm Accessed February 26, 2016

25 Editorial, 'Verdict of Bo Xilai: Life in Prison,' *China Daily*, September 23, 2013, http://www.chinadaily.com.cn/china/2013-09/23/content_16985745.htm Accessed Feb 26 2016

26 Zhang Hong, 'Judge Who Sentenced Bo Xilai to Life in Jail Promoted to Supreme People's Court', *South China Morning Post,* May 21, 2014, http://www.scmp.com/news/china/article/1517282/judge-who-sentenced-bo-xilai-life-jail-promoted-supreme-peoples-court Accessed February 19, 2016

27 Zheng Xuelin, 'Spending Ten Years Polishing a Sword and Showing It Today', January 7, 2015, ref. & trs. Yanmei Lin & Jack Tuholske, 'Field Notes From the Far East: China's New Public Interest Environmental Protection Law', *The Environmental Law Reporter*, vol. 45, no. 9 (September 2015)

28 Lin & Tuholske, op. cit.

29 Yang, op. cit.

30 Xu Nan & Zhang Chun, 'Why Are China's Anti-pollution Lawsuits Stalling?', chinadialogue, June 23, 2015, https://www.chinadialogue.net/article/show/single/en/7986-Why-are-China-s-anti-pollution-lawsuits-stalling- Accessed February 25, 2016

31 ibid.

32 Michael Wilkinson, 'Xi Jinping: China Has "Room for Improvement" on Human Rights', *The Daily Telegraph*, October 21, 2015, http://www.telegraph.co.uk/news/politics/11944803/Xi-Jinpings-UK-visit-to-secure-30bn-Chinese-investment-live.html Accessed September 1, 2016

33 John van der Luit-Drummond, 'European Lawyers Call for Release of China's Human Rights Defenders', *Solicitors Journal*, October 26, 2015, http://www.solicitorsjournal.com/news/public/administrative-and-constitutional/24358/european-lawyers-call-release-china%E2%80%99s-human-righ Accessed February 28, 2016

34 Letters, 'China Must End Its Intimidation and Detention of Human Rights Lawyers', *The Guardian*, January 18, 2016, http://www.theguardian.com/world/2016/jan/18/china-must-end-its-intimidation-and-detention-of-human-rights-lawyers Accessed February 28, 2016

35 CCICED, op. cit.

36 Bie Tao, 'Environment Law System in China', *China Daily*, September 26, 2005, http://www.chinadaily.com.cn/english/doc/2005-09/26/content_480788.htm Accessed February 27, 2016

37 Zhang Chun, 'Growing Pains for China's New Environmental Courts', chinadialogue, June 17, 2015 https://www.chinadialogue.net/article/show/single/en/7972-Growing-

pains-for-China-s-new-environmental-courts Accessed February 27, 2016

38 Cao Yin, 'Top Court Official: Judges Should Take New View of Green Cases', *China Daily*, November 30, 2015, http://www.chinadaily.com.cn/china/2015-11/30/content_22528766.htm Accessed February 27, 2016

10. The Judgement of Paris

1   James Thornton et al., 'Deeds, Not Hot Air', *The Times*, December 4, 2015, full text and signatories at http://www.clientearth.org/open-letter-world-leaders/ Accessed March 23, 2016

2   Quotations and content are from the transcript of a parallel talk James gave to the Environmental Change Institute, University of Oxford, March 1, 2016

3   For an elucidation of this, drawn from the intersection of separate disciplines, see Howard Covington, James Thornton & Cameron Hepburn, 'Global Warming: Shareholders Must Vote for Climate-change Mitigation', *Nature*, vol. 530, no. 7589 (February 11, 2016)

4   Richard Heede, 'Tracing Anthropogenic Carbon Dioxide and Methane Emissions to Fossil Fuel and Cement Producers, 1854–2010', *Climatic Change*, vol. 122, no. 1 (January 2014), p. 237

5   For a wider analysis of the topic, see James Thornton & Howard Covington, 'Climate Change Before the Court', *Nature Geoscience*, vol. 9, no. 1 (January 2016), pp. 3–5

6   For analysis of this issue, see Nikolaos Christidis et al., 'Dramatically Increasing Chance of Extremely Hot Summers Since the 2003 European Heatwave', *Nature Climate Change*, vol. 5, no. 1 (January 2015), pp. 46–50

7   Alice Garton, 'Companies Must Embrace COP Commitments — Government Can't Do It Alone', *Business Green*, February 17, 2016, http://www.businessgreen.com/bg/opinion/2447291/companies-must-embrace-cop-commitments-government-cant-do-it-alone Accessed September 1, 2016

8   https://shareaction.org/

9   For the resolution, rolling out to other companies, see Rio Tinto 2016 notice of annual general meeting, Resolution 17, p. 3, http://www.riotinto.com/documents/RTL_Notice_of_Meeting_2016.pdf Accessed April 13, 2016

10 Helen Wildsmith, 'An "Aiming for A" Update for UKSIF's Ownership Day', Responsible-Investor.com, March 22, 2016, https://www.responsible-investor.com/home/article/helen_wildsmith_an_aiming_for_a_update_ownership/ Accessed April 12, 2016

11 John Elmes, 'Q&A with Howard Covington', *Times Higher Education*, July 9, 2015, https://www.timeshighereducation.com/people/howard-covington-alan-turing-institute-question-and-answer Accessed March 22, 2016

12 NOAA National Centres for Environmental Information, 'Global Analysis — January

2016', February 2016, http://www.ncdc.noaa.gov/sotc/global/201601 Accessed March 23, 2016

13 Chris Mooney, 'Scientists are Floored by What's Happening in the Arctic Right Now', *The Washington Post*, February 18, 2016, https://www.washingtonpost.com/news/energy-environment/wp/2016/02/18/scientists-are-floored-by-whats-happening-in-the-arctic-right-now/ Accessed March 23, 2016

Conclusion (James Thornton)

1 Garrett Hardin, 'The Tragedy of the Commons', *Science*, vol. 162, no. 3859 (December 13, 1968), pp. 1243–1248

2 Jeremy Grantham, from a business bulletin of July 2013, discussed by Nafeez Ahmed, 'Economists Forecast the End of Growth', *The Guardian*, July 19, 2013, https://www.theguardian.com/environment/earth-insight/2013/jul/19/economy-end-growth-resource-scarcity-costs Accessed November 30, 2016

3 Martin Rees, *Our Final Century: Will Civilisation Survive the 21st Century?*, Heinemann, London, 2003

4 E.O. Wilson, *Half-Earth: Our Planet's Fight for Life*, New York, Liveright, 2016

Index